MULTIPLE CHOICE QUESTIONS IN

Human

Anatomy

Ian Whitmore MD, MB BS, LRCP, MRCS
Senior Lecturer in Topographical Anatomy,
Department of Anatomy,
Queen Mary and Westfield College,
London, UK

Peter L.T. Willan MB ChB, FRCS
Senior Lecturer in Anatomy,
Medical School,
University of Manchester,
UK

M Mosby

London Baltimore Bogotá Boston Buenos Aires Caracas Carlsbad, CA Chicago Madrid Mexico City
Milan Naples, FL New York Philadelphia St. Louis Sydney Tokyo Toronto Wiesbaden

Project Manager:	Dave Burin
Developmental Editor:	Lucy Hamilton
Illustration:	Ian Whitmore
Layout:	Jonathan Brenchley
Cover Design:	Ian Spick
Production:	Mike Heath
Publisher:	Fiona Foley

Published in 1995 by Mosby, an imprint of Times Mirror International Publishers Limited

Printed and bound in Great Britain by Biddles Ltd, Guildford and King's Lynn

ISBN 0 7234 2055 6

For full details of all Times Mirror International Publishers Limited titles, please write to Times Mirror International Publishers Limited, Lynton House, 7–12 Tavistock Square, London WC1H 9LB, England.

A CIP catalogue record for this book is available from the British Library.

Library of Congress Cataloging-in-Publication Data has been applied for.

Contents

Preface

This book has been stimulated by the continuing widespread use of multiple choice questions (MCQs) at universities and colleges, both in undergraduate and postgraduate examinations.

The questions in the book are arranged by region, since most courses of human topographical anatomy include the study of cadavers so that structures are considered in their regional context (whether or not the course is 'system based').

We have included some questions that illustrate clinical relevance; both authors know from their personal experience that topographical anatomy underpins clinical examination and diagnosis, as well as some aspects of treatment.

We are aware of the critisism that MCQs tend to encourage undue emphasis on detail, and we have therefore endeavoured to set a range of questions, including some which are basic ('easy') and others which are detailed.

From the wide variety of question formats available, seven different types of question have been selected, including paragraphs for completion and picture questions. Answers and explanatory notes are provided at the end of the book. We hope that the diversity of question types will help maintain the reader's interest in that most fascinating discipline of topographical anatomy.

Although unambiguous and reliable, MCQs are difficult to compose and therefore some institutions endeavour to hoard questions and keep them secret; we would encourage free access to all types of examination questions so that candidates may see what the examiners think is important.

Introduction / User Guide

Specific instructions at the beginning of each group of questions explain how you should answer the different types of question. To give yourself the best chance of success in examinations, you must be familiar not only with the information to be tested, but also with the format of the examination and the marking scheme.

We suggest that you will gain the maximum benefit from these practice questions if you attempt them after you have completed the appropriate part of your study. Answer the questions under conditions which mimic the examination: choose an opportunity when you will not be disturbed, set a time limit, and do not refer to books or notes. Always read the instructions carefully before starting, so that you know what you are required to do, and then read all of the words in each question, looking out for problem words such as 'always' and 'never'. Give clear, accurate, unambiguous answers, and then check through your responses within your time limit to ensure that you have actually written what you intended to write. At the end of the session, mark your responses by comparing them with the answers and explanations provided. Some questions require recall of factual information whereas others require deductive reasoning and are more difficult. Do not be disappointed if you get some answers wrong. Mistakes will identify areas of weakness and will provide a stimulus for further reading to increase your knowledge and understanding.

It is essential that you know whether or not negative marking is in use and that you understand the implications of loosing marks through wrong answers. Marks to be deducted for wrong answers are usually calculated from $(1/(n-1))$, where n is the number of options available. Therefore, a wrong answer to a 'true/false' question scores -1 whereas a wrong answer to a 'one appropriate option from four' question scores $-\frac{1}{3}$. Although the overall result of a large number of random guesses should (statistically) be zero, you are at risk of loosing more marks than you gain. If there is negative marking, we suggest that you answer those questions about which you are knowledgeable and ignore (or indicate by 'don't know') those about which you are totally ignorant. Then review and ponder those questions where you do have some knowledge, since your chances of getting the correct answers for 'educated guesses' will be better than 50:50. However, you should not guess if you do not know answers in an examination which employs negative marking.

It is our experience that many candidates have failed examinations through errors in technique. Therefore, although the comments above have been made many times before, you should reread them several times since you are at risk of making the same mistakes as your predecessors. You should then be able to enter your examinations with confidence that you possess the appropriate examination skills.

1. THORAX

1.1 The internal thoracic artery terminates as musculophrenic and superior epigastric arteries.

1.2 The right pulmonary artery is connected to the aortic arch via the ligamentum arteriosum.

1.3 Branches of the pulmonary artery accompany the divisions of the bronchus in the lung.

1.4 The cardiac plexus receives contributions from the vagus nerve.

1.5 Leakage of blood through the aortic valve during diastole is prevented by chordae tendineae.

1.6 Features visible in the right atrium on the interatrial septum include:
a) the orifice of the coronary sinus.
b) the valve of the inferior vena cava.
c) the fossa ovalis.
d) the crista terminalis.

1.7 In the mediastinum:
a) the left brachiocephalic vein passes behind the left common carotid artery.
b) the brachiocephalic trunk arises from the aortic arch.
c) the left vagus nerve crosses the aortic arch.
d) the ligamentum arteriosum links the aortic arch with the left pulmonary artery.

1.8 Intercostal spaces drain:
a) to the azygos vein on the right posteriorly.
b) via internal thoracic veins to brachiocephalic veins.
c) direct to the brachiocephalic veins for the second and third spaces.
d) to the hemiazygos vein posteriorly on the left.

1.9 The right lung and its pleura:
    a) possess a transverse fissure.
    b) are in contact with the pericardium overlying the right ventricle.
    c) possess an oblique fissure separating the lower from the middle lobe.
    d) have an impression from the azygos arch on their medial surface.

Each of the incomplete statements below is followed by five suggested answers or completions. Decide which are true and which are false.

1.10 A typical rib:
    a) articulates with the transverse process of the thoracic vertebra of the same number.
    b) possesses a head which articulates with the body of the same numbered vertebra.
    c) is attached by a costal cartilage to the sternum.
    d) is attached to the rib below by fibres of external intercostal muscle.
    e) has visceral pleura in contact with its deep surface.

1.11 The left lung:
    a) has three lobes.
    b) is grooved by the aortic arch.
    c) contains 10 bronchopulmonary segments.
    d) is in contact with the heart.
    e) possesses a lingula.

1.12 The oesophagus:
    a) passes through the right crus of the diaphragm.
    b) receives innervation from the phrenic nerve.
    c) is indented by the arch of the aorta.
    d) is closely related to the right recurrent laryngeal nerve in the thorax.
    e) is in contact with the anterior surface of the trachea.

1.13 Sounds from the following intrathoracic structures are best heard with a stethoscope at the paired positions:
    a) middle lobe of right lung – lateral surface of thorax just below right axilla.
    b) apex of upper lobe of left lung – just below lateral third of left clavicle.
    c) tricuspid valve – 5th left intercostal space midclavicular line.
    d) aortic valve – anterior end of right 2nd intercostal space.
    e) mitral valve – xiphisternum.

1.14 A typical thoracic vertebra:
  a) possesses a transverse process that articulates with the head of a rib.
  b) articulates with the body of the vertebra above it at a synovial joint.
  c) has the spinal nerve of the same number emerging above it.
  d) possesses foramina transversaria.
  e) possesses a spinous process projecting horizontally and posteriorly.

1.15 The left border of the mediastinal shadow in a PA radiograph includes:
  a) the left ventricle.
  b) the right ventricle.
  c) the left atrial appendage.
  d) the pulmonary trunk.
  e) the aortic arch.

1.16 The trachea:
  a) has the right brachiocephalic vein anteriorly.
  b) divides at the level of the fourth thoracic vetrebra.
  c) has the aortic arch on its left.
  d) has a sensory supply from the phrenic nerves.
  e) is closely related to the recurrent laryngeal nerves.

Each question consists of two statements. Choose:
a) if both statements are true and they are causally related.
b) if both statements are true but they are not causally related.
c) if the first statement is true and the second is false.
d) if the first statement is false and the second is true.
e) if both statements are false.

1.17 Pain is felt at the shoulder tip with diaphragmatic irritation
  because
  the dermatome for C4 is at the shoulder tip and C4 fibres are contained in the phrenic nerve.

1.18 The mitral and tricuspid valves open during ventricular diastole
  because
  the papillary muscles tense chordae tendineae during contraction.

1.19 Inhaled particles are most likely to enter the lower lobe of the left lung
  because
  the left main bronchus is more vertical then the right.

1.20 Carcinoma of one breast may spread to the other breast
because
the lymphatic drainage from the breast is mainly to the axillary lymph nodes.

1.21 Lymph drains from the abdomen to the junction of the brachiocephalic veins
because
the thoracic duct passess through the posterior mediastinum.

1.22 Pain is felt from the pleura in penetrating injuries entering the lung
because
parietal pleura has a sensory nerve supply from the overlying intercostal nerve.

1.23 Needle biopsy of the liver through the costodiaphragmatic recess avoids the lung
because
the lung never occupies this recess.

1.24 Admitting air into the space between the two layers of pleura surrounding the left lung will lead to collapse of the right lung
because
there is communication between the pleural spaces surrounding both lungs.

In the following text, some words or phrases have been replaced by letters in brackets. Select the most appropriate word or phrase for each letter. Where a letter appears more than once, it represents exactly the same word or phrase.

1.25 The left coronary artery originates from the (a)_____ just above the (b)_____ valve. The left coronary artery divides into the (c)_____ which passes down the (d)_____ groove to the apex of the heart, and the (e)_____ which traverses the (f)_____ groove to the posterior surface of the heart accompanied by the (g)_____. The (g) is formed by the union of the (h)_____ and (j)_____ and drains into the (k)_____. The right coronary artery passes down the (f) groove giving branches to the (l)_____ atrium. The sinuatrial node is usually supplied by the (m)_____ coronary artery and the atrioventricular node by the (n)_____ coronary artery.

1.26 A typical thoracic spinal nerve arises from the spinal cord by a (a)_____ which possesses a swelling, known as the (b)_____, and by a (c)_____. The (a) and (c) unite to form the spinal nerve which divides into a (d)_____ which supplies skin and muscle near the midline posteriorly, and a (e)_____. Sympathetic axons leave the (e) via a (f)_____ which connects to the sympathetic trunk. Axons leaving the trunk to return to the (e) are (g)_____ and form a (h)_____. The area of skin supplied by (d) and (e) is known as a (j)_____ and the muscles constitute a (k)_____.

1.27 The phrenic nerve arises in the neck from the anterior rami of (a)_____ and enters the thorax through the thoracic (b)_____ behind the subclavian (c)_____. The right phrenic nerve passes down the lateral edge of the mediastinum covered by (d)_____ and in contact with the (e)_____and the (f)_____, before coming into contact with the pericardium covering the (g)_____. The left phrenic nerve also passes down the lateral border of the mediastinum, but on the left, passing over the (h)_____ and then the (j)_____. It continues over the pericardium covering the (k)_____, the (l)_____ and the (m)_____ before piercing the diaphragm. The phrenic nerves convey sensation from (n)_____ and _____ pleura, (o)_____ and_____.

1.28 In the conducting system of the heart, impulses originate at the (a)_____ which lies in the (b)_____ wall of the (c)_____ atrium close to the termination of the (d)_____ at the upper end of the (e)_____. They are propagated across the myocardium of the atria to the (f)_____, which is situated in the (g)_____ septum, anterosuperior to the opening of the (h)_____. From (f), exitation waves pass down the (j)_____ to right and left (k)_____es. The right (k) often gives a branch that crosses from the interventricular septum to the anterior wall of the right ventricle in the (l)_____.

1.29 Identify the indicated structures in this drawing of a magnetic resonance image of the thorax.

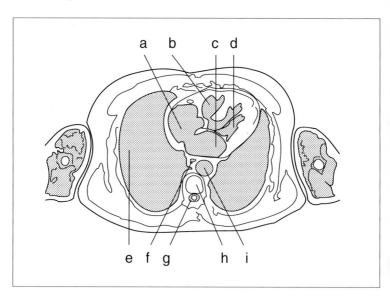

2. UPPER LIMB

Indicate whether the statement is true or false.

2.1 Extension of the interphalangeal joint of the thumb is produced by both extensor pollicis longus and extensor pollicis brevis.

2.2 Lateral rotation of the shoulder joint is produced by infraspinatus.

2.3 Movements in both the coronal and the sagittal planes are possible at the wrist (radiocarpal) joint.

2.4 The axillary nerve accompanies the profunda brachii artery for part of its course.

2.5 Movement at the sternoclavicular joint contributes significantly to full abduction of the upper limb.

2.6 Teres minor separates the intermuscular spaces through which the axillary and the radial nerves leave the axilla.

2.7 Latissimus dorsi is supplied by branches of the intercostal nerves.

2.8 Pectoralis minor is an accessory muscle of ventilation.

2.9 The thoracodorsal nerve supplies skin of the posterolateral aspect of the chest wall.

2.10 The acromioclavicular joint provides the only articulation between the axial skeleton and the upper limb.

2.11 Serratus anterior protracts the scapula.

2.12 Pectoralis major is a flexor of the shoulder joint.

2.13 The carpal tunnel is bounded anteriorly by the palmar aponeurosis.

2.14 The palmar aponeurosis attaches distally to the digital fibrous flexor sheaths.

2.15 The median nerve accompanies the brachial artery throughout the artery's entire length.

2.16 The proximal part of the tendon of the long head of biceps brachii is intracapsular.

2.17 Part of triceps attaches to the capsule of the elbow joint.

2.18 Brachioradialis is supplied by the radial nerve.

2.19 The principal sensory nerves supplying the thumb are branches of the radial nerve.

2.20 The anterior interosseous nerve supplies flexor digitorum profundus.

2.21 The long thoracic nerve is a branch of the posterior cord of the brachial plexus.

2.22 The ulnar nerve arises directly from the lower trunk of the brachial plexus.

2.23 Skin of the thenar eminence is supplied by branches of the median nerve.

2.24 Serious trauma which tears the roots and trunks of the brachial plexus results in paralysis of trapezius.

2.25 The characteristic rounded contour of the shoulder region is due to deltoid overlying the proximal end of the humerus.

Each of the incomplete statements below is followed by four or five suggested answers or completions. Select the one that is most appropriate.

2.26 Lateral (external) rotation of the humerus at the glenohumeral joint is produced by:
   a) supraspinatus.
   b) deltoid.
   c) pectoralis major.
   d) teres major.
   e) latissimus dorsi.

2.27 Which of the following muscles produce adduction of the fingers?
a) dorsal interossei.
b) palmar interossei.
c) lumbricals.
d) extensor digitorum.

2.28 The following are true of the metacarpal bone of the thumb, *except*:
a) it receives the tendon of abductor pollicis longus.
b) it articulates with sesamoid bones .
c) it receives attachment of abductor pollicis brevis.
d) it articulates with the proximal phalanx at a hinge joint.
e) it articulates with the trapezium.

2.29 The following are true of the scaphoid, *except*:
a) it is palpable only in the anatomical snuff box.
b) it has several articular facets covered with cartilage.
c) it has no muscle or tendon attachments.
d) it articulates with the distal surface of the radius.
e) it transmits body weight in a fall on the hand.

2.30 The following muscle is not attached to the radius:
a) extensor pollicis brevis.
b) abductor pollicis longus.
c) extensor pollicis longus.
d) biceps brachii.
e) flexor pollicis longus.

2.31 In a fall on the hand, direct transmission of body weight involves the following bone, which is at risk of fracture:
a) the olecranon process.
b) the head of the radius.
c) the distal end of the ulna.
d) the pisiform.
e) the clavicle.

2.32 Which of these muscles does not form part of the 'rotator cuff'?
a) supraspinatus.
b) infraspinatus.
c) teres minor.
d) subscapularis.
e) teres major.

2.33 Inability to extend the distal interphalangeal joint of the index finger is caused by complete division of the:
a) deep branch of the ulnar nerve.
b) radial nerve at the level of the wrist joint.
c) tendon of extensor indicis.
d) median nerve at the level of the wrist joint.
e) radial nerve at the level of the elbow joint.

2.34 The following muscles attach to the humerus, *except*:
a) pectoralis major.
b) biceps brachii.
c) brachialis.
d) coracobrachialis.
e) subscapularis.

2.35 Attached to the lesser tubercle of the humerus is:
a) teres minor.
b) teres major.
c) subscapularis.
d) coracobrachialis.
e) pectoralis major.

2.36 Which of the following muscles does not attach to the scapula?
a) pectoralis minor.
b) triceps brachii.
c) biceps brachii.
d) brachialis.
e) serratus anterior.

2.37 The following lie in the carpal tunnel:
a) the radial artery.
b) the ulnar nerve.
c) the median nerve.
d) the radial nerve.
e) the ulnar artery.

2.38 Which of these spinal cord segments does not supply skin of the digits?
a) T1.
b) C6.
c) C8.
d) C7.

2.39 The lateral cord of the brachial plexus:
 a) includes fibres from spinal nerve C8.
 b) supplies deltoid muscle.
 c) provides fibres which enter the lateral cutaneous nerve of the forearm.
 d) receives fibres from the same segments as the posterior cord.

2.40 The anterior interosseous nerve of the forearm:
 a) supplies all the deep muscles of the forearm.
 b) is derived partly from the ulnar nerve.
 c) is a purely motor nerve.
 d) lies deep to flexor digitorum superficialis.

2.41 A birth injury to the upper parts of the brachial plexus would be least likely to involve:
 a) pectoralis major.
 b) the thenar muscles.
 c) biceps brachii.
 d) deltoid.

2.42 The posterior interosseous nerve of the forearm does not supply these muscles:
 a) extensor digitorum.
 b) extensor carpi ulnaris.
 c) extensor carpi radialis longus.
 d) extensor digiti minimi.

2.43 The major contributor to stability of the glenohumeral joint is:
 a) the rotator cuff.
 b) synovial membrane.
 c) the glenoid labrum.
 d) the tendon of the long head of biceps.
 e) the coracoacromial arch.

2.44 In the fingers, hand and distal forearm, synovial sheaths associated with tendons:
 a) do not extend distally beyond the proximal interphalangeal joints.
 b) are separate for flexor digitorum profundus and flexor digitorum superficialis.
 c) provide a pathway for spread of infection from the digits to the forearm.
 d) allow communication between the index and middle fingers.

2.45 Complete division of the ulnar nerve at the level of the wrist results in paralysis of:
a) the thenar muscles.
b) all of the lumbrical muscles.
c) the hypothenar muscles.
d) flexor carpi ulnaris.

2.46 The palmar aponeurosis:
a) lies superficial to the median nerve.
b) is crossed superficially by the superficial palmar arch.
c) is deficient when palmaris longus is absent.
d) extends laterally to cover the muscles of the thenar eminence.
e) allows freedom of mobility for the skin of the palm.

2.47 Damage to the ulnar nerve at the level of the midshaft of the humerus will be associated with:
a) loss of flexion of the ring finger.
b) anaesthesia of the palm.
c) wasting of the thenar muscles.
d) weakness of adduction of the thumb.

2.48 Which muscle acting as a prime mover flexes the distal and proximal interphalangeal joints?
a) flexor digitorum superficialis.
b) flexor digitorum profundus.
c) a lumbrical.
d) a dorsal interosseous.
e) a palmar interosseous.

2.49 Which of the following types of nerve fibres are present in the brachial plexus?
a) preganglionic sympathetic fibres.
b) parasympathetic fibres.
c) fibres from anterior rami of spinal nerves.
d) fibres from posterior rami of spinal nerves.

Each of the incomplete statements below is followed by four or five suggested answers or completions. Decide which are true and which are false.

2.50 The flexor retinaculum attaches to:
   a) the styloid process of the ulna.
   b) pisiform.
   c) hamate.
   d) scaphoid.
   e) the fifth metacarpal.

2.51 The following muscles act on the metacarpophalangeal joint of the thumb:
   a) opponens pollicis.
   b) abductor pollicis brevis.
   c) extensor pollicis brevis.
   d) the first dorsal interosseous.
   e) flexor pollicis brevis.

2.52 The extensor expansion of the index finger receives attachment(s) from the following muscles:
   a) the third palmar interosseous.
   b) the first dorsal interosseous.
   c) the second dorsal interosseous.
   d) the first lumbrical.
   e) the second lumbrical.

2.53 Deltoid:
   a) is innervated by fibres from the posterior cord of the brachial plexus.
   b) is usually active against gravity during adduction of the shoulder joint.
   c) cooperates with supraspinatus during abduction of the shoulder joint.
   d) may suffer damage to its nerve supply during dislocation of the glenohumeral joint.
   e) acts as a prime mover during extension of the glenohumeral joint.

2.54 The following pass through the anterior wall of the axilla:
a) the cephalic vein.
b) the tendon of the short head of biceps.
c) coracobrachialis.
d) branches of the thoracoacromial artery.
e) lymphatic vessels.

2.55 The following muscles act as prime movers:
a) pectoralis major during flexion at the shoulder.
b) brachialis during flexion at the elbow.
c) biceps brachii during supination.
d) pronator teres during pronation.
e) trapezius during abduction at the glenohumerus joint.

2.56 Shrugging the shoulders against resistance demonstrates activity in:
a) the spinal accessory nerve.
b) trapezius.
c) the posterior rami of thoracic spinal nerves.
d) sternocleidomastoid.
e) pectoralis major.

2.57 A patient with damage to his musculocutaneous nerve would lack the following active movements:
a) flexion at the shoulder.
b) flexion at the elbow.
c) supination.
d) adduction at the shoulder.

2.58 The following muscles are important in full abduction of the upper limb at the shoulder:
a) serratus anterior.
b) trapezius.
c) deltoid.
d) supraspinatus.
e) rhomboid major.

2.59 Palmaris longus:
   a) is not functionally important.
   b) lies superficial to the median nerve.
   c) attaches to the palmar aponeurosis.
   d) lies deep to the flexor retinaculum.
   e) is innervated by the ulnar nerve.

2.60 Medial rotation of the humerus at the shoulder joint is produced by:
   a) teres minor.
   b) supraspinatus.
   c) deltoid.
   d) latissimus dorsi.
   e) pectoralis major.

2.61 The following enter and/or leave the cubital fossa:
   a) the tendon of biceps brachii.
   b) the superficial branch of the radial nerve.
   c) the cephalic vein.
   d) the ulnar nerve.
   e) the radial artery.

2.62 Collateral vessels permit significant blood flow when there is blockage of the:
   a) radial artery.
   b) ulnar artery.
   c) brachial artery.
   d) axillary artery.
   e) subclavian artery.

2.63 For part of its course, the brachial artery accompanies the:
   a) axillary nerve.
   b) median nerve.
   c) radial nerve.
   d) ulnar nerve.
   e) musculocutaneous nerve.

2.64 The following receive fibres from the lateral cord of the brachial plexus:
a) the median nerve.
b) the axillary nerve.
c) the musculocutaneous nerve.
d) the lateral cutaneous nerve of the forearm.
e) the ulnar nerve.

2.65 Branches of the radial nerve supply:
a) skin of the arm.
b) skin of the forearm.
c) skin of the hand.
d) no muscle that has flexor actions.
e) no muscle(s) located in the hand.

2.66 The following are derived from anterior divisions of the trunks of the brachial plexus:
a) the axillary nerve.
b) the radial nerve.
c) the thoracodorsal nerve.
d) the suprascapular nerve.
e) the long thoracic nerve.

2.67 The following contribute to stability of the proximal radioulnar joint:
a) the anular ligament.
b) triceps brachii.
c) the radioulnar interosseous ligament.
d) the radial collateral ligament of the elbow joint.
e) anconeus.

2.68 Deep (investing) fascia of the forearm:
a) fuses with the aponeurosis of biceps.
b) attaches to the radius.
c) is perforated by the superficial branch of the radial nerve.
d) is perforated by the cephalic vein.
e) provides attachment for superficial muscles in the flexor compartment.

2.69 The long thoracic nerve:
a) supplies skin of the chest wall.
b) lies on the medial wall of the axilla.
c) innervates serratus anterior.
d) arises from the lower trunk of the brachial plexus.
e) is accompanied by the long thoracic artery.

2.70 The following muscles attach to the scapula, humerus and radius:
a) triceps brachii.
b) biceps brachii.
c) coracobrachialis.
d) brachioradialis.
e) brachialis.

2.71 Skin of the dorsum of the hand is innervated by the:
a) lateral cutaneous nerve of the forearm.
b) superficial branch of the radial nerve.
c) ulnar nerve.
d) median nerve.
e) posterior cutaneous nerve of the forearm.

2.72 The following are hinge joints:
a) the elbow joint.
b) the radiocarpal joint.
c) the metacarpophalangeal joint of the thumb.
d) the metacarpophalangeal joint of the little finger.
e) the midcarpal joint.

2.73 The apical group of axillary lymph nodes receives lymph from the:
a) scapular region.
b) breast.
c) anterior abdominal wall.
d) thumb.
e) palm of the hand.

2.74 Asymmetry of the shoulder region would result from unilateral damage to the:
a) spinal accessory nerve.
b) long thoracic nerve.
c) axillary nerve.
d) radial nerve.
e) thoracodorsal nerve.

2.75 In dislocation of the shoulder joint it is appropriate to test cutaneous sensation over the insertion of deltoid
because
the lower lateral cutaneous nerve of the arm is a branch of the radial nerve.

2.76 Weakness of the extensors of the wrist joint may complicate fracture of the humerus near the middle of its shaft
because
in the spiral groove, the radial nerve is closely related to the humerus and is the only nerve to supply the extensor muscles of the wrist joint.

2.77 Extension at the shoulder joint is impaired if the radial nerve is damaged
because
the radial nerve supplies the extensor compartment of the arm.

2.78 Biceps brachii acts on both the humeroulnar and the proximal radioulnar joints
because
biceps attaches distally only to the ulna.

2.79 Deltoid is an abductor at the glenohumeral joint
because
that part of deltoid which attaches to the acromion has a multipennate arrangement of its muscle fibres.

2.80 Fracture of the humerus near the middle of its shaft is associated with clawing of the hand
because
the radial nerve may be damaged near the midshaft of the humerus, resulting in impaired grip.

2.81 Congenital absence of part of pectoralis major is associated with serious limitation of every day activities
because
other muscles cannot compensate for the absence of pectoralis major in activities which require power.

2.82 Following fracture through the surgical neck of the humerus, the short proximal part of the bone is usually adducted
because
spasm of pectoralis major and latissimus dorsi keeps the bone in an adducted position.

2.83 Where skin is tethered near a joint, the skin creases lie at right angles to the axis of movement
because
the muscles which produce the movement have tendinous attachments to the dermis of the skin.

2.84 In the carpal tunnel syndrome, sensation in most of the skin of the palm is intact
because
the median nerve does not traverse the carpal tunnel.

2.85 An ulnar nerve lesion is associated with hyperextension of the meta-carpophalangeal joints
because
at these joints the extensor digitorum muscle is no longer opposed by the interosseous muscles.

2.86 When the carpal tunnel syndrome is suspected it is useful to examine adductor pollicis
because
the belly of the muscle is easy to palpate.

2.87 Flexor digitorum superficialis is innervated by the anterior interosseous nerve
because
the nerve supplies flexor muscles of the forearm.

2.88 The distal end of the ulna does not articulate with the triquetral
because
a triangular cartilage attaches the radius to the styloid process of the ulna.

2.89 Flexor digitorum profundus has no action on the distal interphalangeal joints
because
its tendon lies deep to the flexor digitorum superficialis and attaches to the middle phalanges.

2.90 Opposition of the fingers is impossible
because
the carpometacarpal joints of the fingers do not allow rotation.

2.91 Pronator teres is not a powerful flexor of the elbow joint
because
it is the most laterally placed of the superficial flexor muscles.

2.92 Abductor pollicis longus does not assist extension of the thumb
because
the muscle is not a component of the extensor compartment of the forearm.

2.93 The radial nerve contains nerve fibres from spinal nerves C6, C7 and C8
because
the lateral and medial cords of the brachial plexus both receive contributions from C6, C7, and C8.

2.94 Damage to the radial nerve does not impair grip
because
the flexors of the fingers and thumb are supplied by the median and ulnar nerves.

2.95 Injury to the musculocutaneous nerve weakens supination of the forearm
because
brachialis is innervated by the musculocutaneous nerve.

2.96 The interosseous membrane resists distal displacement of the radius in relation to the ulna
because
the fibres in the membrane are directed inferiorly and laterally.

2.97 Brachioradialis contributes to rotation of the forearm
because
its distal and proximal attachments lie on the same sagittal plane.

2.98 The superficial palmar arch receives a contribution from the radial artery
because
near the wrist, the radial artery crosses onto the posterior aspect of the hand.

2.99 Trapezium and trapezoid do not articulate with the radius
because
both bones are located on the medial aspect of the carpus.

In the following text, some words or phrases have been replaced by letters in brackets. Select the most appropriate word or phrase for each letter. Where a letter appears more than once, it represents exactly the same word or phrase.

2.100 Biceps brachii is located in the (a)_____ compartment of the (b)_____. As its name implies, biceps possesses (c)_____, the (d)_____ lying medially and the (e)_____ located laterally. The (d) attaches with (f)_____ to the coracoid process, whereas the (e) attaches to the scapula at its (g)_____ tubercle. The (e) attaches by means of a tendon which passes (h)_____ to the (i)_____ ligament lying across the intertubercular groove and possesses a sheath of (j)_____ within the (k)_____ joint. Distally, the muscle attaches to the (l)_____ border of the ulna by an (m)_____ which reinforces the roof of the (n)_____ fossa and to the (o)_____ by means of a substantial tendon which enters the (n) fossa. Biceps is supplied by the (p)_____ nerve. In addition to acting as a powerful flexor of the (q)_____ joint and a weak flexor of the (r)_____ joint, biceps acts on the (s)_____ joints being an important (t)_____ of the forearm, particularly when the elbow is in a position of (u)_____.

2.101 The cephalic vein arises from the (a)_____ side of the (b)_____ on the (c)_____ aspect of the hand. The vein crosses the anatomical (d)_____ where it is closely related to branches of the (e)_____ nerve. At the level of the (f)_____ joint the vein lies (g)_____ in position and frequently communicates with the (h)_____ vein on the medial aspect of the limb. The cephalic vein ascends in the (i)_____ fascia lateral to (j)_____ muscle and more proximally enters the groove between (k)_____ laterally and (l)_____ medially. The vein turns deeply (m)_____ to the clavicle and terminates in the (n)_____ vein. The cephalic vein is accompanied by (o)_____ which drain the superficial tissues on the (p)_____ side of the limb to lymph nodes in the (q)_____ fossa or to deep lymph nodes within the (r)_____.

2.102 Identify the parts labelled a–i.

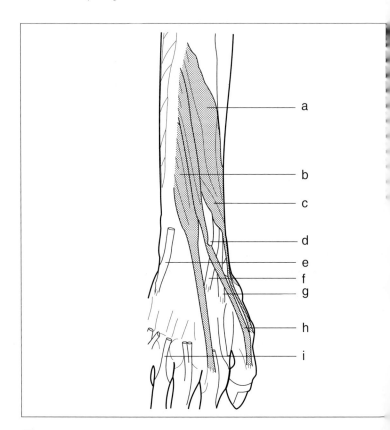

2.103 Pair each of the labelled parts a–i (see **Question 2.102**) with the most appropriate from the list of items A–I using each only once.

A acts on interphalangeal joint of thumb.
B lateral border of anatomical snuff box.
C attaches to third metacarpal bone.
D 'pulley' for tendon of extensor pollicis longus.
E acts on metacarpal of thumb but not its phalanges.
F attaches to lateral supracondylar ridge of humerus.
G extensor of ring finger.
H adductor of wrist joint.
I attaches to dorsal surface of ulna and interosseous membrane.

2.104 Identify the parts labelled a–i.

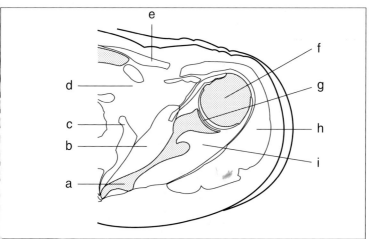

2.105 Pair each of the labelled parts a–i (see **Question 2.104**) with the most appropriate from the list of items A–I using each only once.

A concave articular surface.
B fluid at body temperature.
C thin flat bone.
D protracts the scapula.
E medial rotator of shoulder joint.
F flexor of shoulder joint.
G supplied by axillary nerve.
H lateral rotator of shoulder joint.
I limited by anatomical neck.

3. ABDOMEN

Indicate whether the statement is true or false.

3.1 The head of the pancreas is supplied by branches of the superior mesenteric artery.

3.2 The bile duct enters the first part of the duodenum.

3.3 The subcostal nerve lies anterior to the left kidney.

3.4 The left suprarenal gland drains blood to the left renal vein.

3.5 The lesser omentum attaches to the liver at the porta hepatis.

3.6 The cystic duct joins the left hepatic duct to form the bile duct.

3.7 The taeniae coli meet at the attachment of the ileum to the caecum.

Each of the incomplete statements below is followed by four suggested answers or completions. Select the one that is incorrect.

3.8 The liver:
a) lies deep to the 10th intercostal space.
b) lies across the midline.
c) receives blood from the small intestine via the hepatic portal vein.
d) is in direct contact with the left dome of the diaphragm.

3.9 The following structures lie posterior to the left kidney:
a) the subcostal nerve.
b) the 12th rib.
c) psoas major.
d) perinephric fat.

3.10 In the anterior abdominal wall:
   a) pain from the umbilical area passes in the 10th intercostal nerve.
   b) the linea semilunaris crosses the costal margin at the seventh costal cartilage.
   c) the rectus abdominis attaches to the posterior surface of the costal margin.
   d) the linea alba separates the two rectus abdominis muscles.
   e) the anterior rectus sheath contains the external oblique aponeurosis.

3.11 The gall bladder:
   a) lies behind the tip of the seventh costal cartilage.
   b) concentrates and stores bile.
   c) possesses a spiral valve.
   d) is supplied by the cystic artery.
   e) is attached to the visceral surface of the liver by peritoneum.

3.12 The caecum:
   a) receives its blood supply from the inferior mesenteric artery.
   b) is retroperitoneal.
   c) lies in the right iliac fossa.
   d) possesses taeniae coli.
   e) has the vermiform appendix attached to its lateral wall.

3.13 Concerning the small intestine:
   a) jejunal arterial arcades are simpler than ileal arcades.
   b) the first part of the duodenum is supplied by the superior mesenteric artery.
   c) it is all intraperitoneal.
   d) the mucosa is raised into plicae circulares.
   e) lymph drains into the cysterna chyli before leaving the abdomen.

3.14 The following are intraperitoneal:
   a) the ascending colon.
   b) the transverse colon.
   c) the stomach.
   d) the head of the pancreas.
   e) the spleen.

3.15 The inferior mesenteric artery:
a) passes to the right of the midline.
b) supplies the sigmoid colon.
c) is wholly contained in the abdomen.
d) gives a middle colic branch.
e) supplies the descending colon.

3.16 The left renal vein is more likely to suffer thrombosis than the right
because
the left renal vein is longer and is frequently compressed as it passes
across the front of the abdominal aorta.

3.17 Blood from the left gonad drains into the portal venous system
because
the left gonadal vein drains into the left renal vein.

3.18 In a paramedian incision through the rectus sheath, the rectus abdominis
muscle should be displaced laterally
because
the nerve supply to rectus abdominis enters on its lateral side.

3.19 The most superior midline branch of the abdominal aorta is the
(a)_____, which is surrounded by elements of the autonomic nervous
system known as (b)_____. The sympathetic input to (b) is from
(c)_____, which traverse the diaphragm and originate from the thoracic
(d)_____. The fibres in (c) are mostly (e)_____. A branch of (a), the

(f)_____ passes to the cardia and gives rise to ascending (g)_____ branches. Another branch of (a), the (h)_____ also passes to the left, closely related to the pancreas, giving a few (j)_____ branches and the left (k)_____ just before terminating in the (l)_____.

3.20 The stomach possesses a superior part, the (a)_____, which contains (b)_____ when the patient is upright. At the right extremity of (a), the oesophagus enters at the (c)_____. Continuing inferiorly from (c) along the (d)_____ is the (e)_____ artery and vein. The (e) vein drains into the (f)_____ and communicates with veins in the thorax via (g)_____ veins. This junction between the two systems is known as a (h)_____ and is a frequent site of (j)_____ which can bleed profusely. On the anterior surface of the oesophagus is a branch of the vagus nerve, the (k)_____ which is derived mainly from the (l)_____ vagus.

3.21 Identify the indicated structures in this drawing of a cross section of the abdomen seen from above:

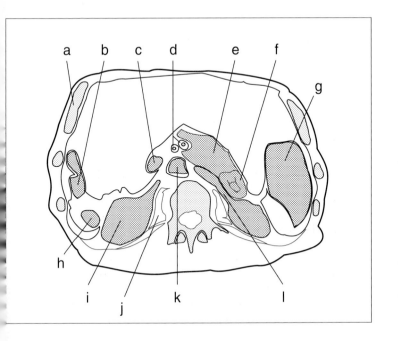

4. PELVIS AND PERINEUM

Indicate whether the statement is true or false.

4.1 The narrowest part of the male urethra is located where it traverses the perineal membrane.

4.2 Venous drainage from the testis passes to the inferior mesenteric vein.

4.3 Lymph from the left ovary passes to paraaortic nodes.

4.4 The uterine artery crosses the ureter just lateral to the cervix.

4.5 The nerves of the uterus are derived from the inferior hypogastric plexus.

4.6 The perineal body lies anterior to the vagina.

Each of the incomplete statements below is followed by four suggested answers or completions. Select the one that is incorrect.

4.7 The parasympathetic nerves of the pelvis:
a) arise from S2, S3 & S4.
b) ascend through the hypogastric plexus to supply the sigmoid colon.
c) form part of the micturition reflex.
d) carry the nervous impulse responsible for ejaculation.

4.8 The corpus spongiosum:
a) contains the penile urethra.
b) attaches to the superior surface of the perineal membrane.
c) is compressed by bulbospongiosus.
d) receives blood from the deep artery of the penis.

4.9 The greater vestibular gland:
a) lies in the deep perineal pouch.
b) has a duct which opens into the vestibule of the vagina.
c) is closely related to the bulb of the vestibule.
d) is covered by bulbospongiosus.

4.10 Sphincter urethrae externa:
  a) lies in the deep perineal pouch.
  b) is pierced by the urethra.
  c) receives sympathetic fibres from S2, S3 & S4.
  d) is under voluntary control.

4.11 Rupture of the penile bulbar urethra gives rise to extravasation of urine into:
  a) the superficial fascia of the scrotum.
  b) the superficial fascia of the penis.
  c) the superficial fascia of the thigh.
  d) the superficial fascia of the lower anterior abdominal wall.

Each of the incomplete statements below is followed by five suggested answers or completions. Decide which are true and which are false.

4.12 The peritoneum in the pelvis:
  a) forms the mesovarium.
  b) is raised into a fold by the ductus deferens.
  c) has its most inferior point between bladder and uterus in the female.
  d) has its most inferior point between bladder and rectum in the male.
  e) is in contact with the ureter.

4.13 The uterus:
  a) has a lumen which communicates with the peritoneal cavity.
  b) derives its arterial supply from the external iliac artery.
  c) is closely related to the anal canal.
  d) is supported in the broad ligament.
  e) has lymphatic drainage primarily to inguinal nodes.

4.14 The testis:
  a) has lymphatic drainage to the inguinal lymph nodes.
  b) receives its arterial blood supply from the internal pudendal artery.
  c) is elevated by the striated muscle in the cremaster muscle.
  d) is maintained at a constant lower than body core temperature by the action of the cremaster muscle.
  e) has venous drainage on the left to the left renal vein.

4.15 The cervix uteri:
   a) has the ureter as a close relation.
   b) is the most mobile part of the uterus.
   c) possesses an external os on its inferior surface.
   d) is closely related to the uterine artery laterally.
   e) lies completely within the vagina.

4.16 The right ureter in the female:
   a) opens into the base of the bladder.
   b) crosses the bifurcation of the common iliac artery.
   c) passes through the levator ani.
   d) lies on psoas major muscle.
   e) receives blood from the right renal artery.

4.17 The prostate:
   a) receives the secretions of the seminal vesicles through the ejaculatory ducts.
   b) possesses a median lobe.
   c) is related anteriorly to peritoneum.
   d) has the ductus deferens on its anterior wall .
   e) lies inferior to the bladder.

4.18 The vagina:
   a) possesses an anterior wall pierced by the cervix uteri.
   b) lies anterior to the urethra.
   c) is separated from the bladder by peritoneum.
   d) forms an angle of 90° or more with the uterus.
   e) is closely related to the ureter.

Each question consists of two statements. Choose:
   a) if both statements are true and they are causally related.
   b) if both statements are true but they are not causally related.
   c) if the first statement is true and the second is false.
   d) if the first statement is false and the second is true.
   e) if both statements are false.

4.19 Pain in appendicitis may be felt in the right knee
   because
   referred pain is carried in the obturator nerve following irritation where it crosses the pelvic wall.

4.20 The levator ani muscles provides important support for the uterus
because
the levator ani relaxes during coughing and sneezing.

4.21 The cervix uteri cannot be palpated through the rectum
because
it is separated from the rectum by the rectouterine pouch.

4.22 Infection in the ischiorectal (ischioanal) fossa may spread to the fossa on the other side
because
the two fossae are continuous across the midline behind the anal canal.

4.23 Pudendal nerve block at the ischial spine does not provides anaesthesia of the labia majora
because
the perineal branch of the pudendal nerve begins in the gluteal region.

4.24 Pus in the rectouterine pouch often drains via the vagina
because
the posterior fornix of the vagina is closely related to the rectouterine pouch by a thin layer of peritoneum.

4.25 The bladder enlarges in an inferior direction
because
the trigone is superior.

In the following text, some words or phrases have been replaced by letters in brackets. Select the most appropriate word or phrase for each letter. Where a letter appears more than once, it represents exactly the same word or phrase.

4.26 The levator ani muscle takes origin from the (a)_____ bone anteriorly and the (b), which is a thickening in the (c)_____ fascia overlying (d)_____ muscle. Attachment of the (b) extends posteriorly as far as the (e)_____ and the general direction of levator ani fibres from origin to insertion is (f)_____. The motor nerve supply is derived from (g)_____ and (h)_____ and contraction results in (i)_____ of the pelvic contents and (j)_____ of the anorectal junction. The muscle forms the superior border of a fat-filled space, the (k)_____.

4.27 The common iliac artery divides into the more lateral (a)_____ and the (b)_____. This bifurcation is crossed by the (c)_____ and in the female the (d)_____ is anteroinferior. The (b) gives rise to an anterior division with a (e)_____ branch to the bladder and the (f)_____ to the lower limb. More medially placed branches include the (g)_____ to the perineum, the (h)_____ to the rectum and in the female (i)_____ and (j)_____ branches. The posterior division of the (b) gives rise to the (k)_____ which leaves the pelvis through the (l)_____ to supply structures in the buttock. Also passing through the (l) from the anterior division is the (m)_____.

4.28 Identify the indicated structures in this drawing of a dissection of the right half of the female pelvis seen from the midline:

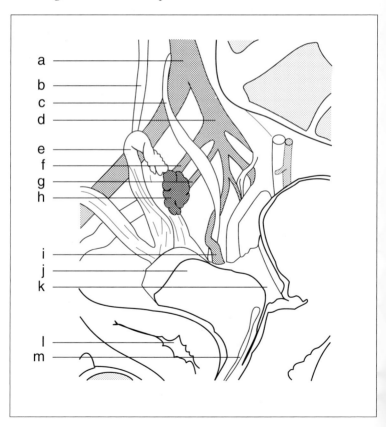

5. LOWER LIMB

5.1 The direction of pull of rectus femoris muscle tends to prevent dislocation of the patella.

5.2 Quadriceps is active during standing to stabilise the knee joint.

5.3 The popliteal fascia is perforated by the great saphenous vein.

5.4 The hip joint is extended by adductor magnus.

5.5 The deep peroneal nerve enters the foot anterior to the ankle joint.

5.6 Adductor magnus is supplied by the obturator nerve alone.

5.7 Spinal nerves L3 and L4 supply skin over the medial side of the leg.

5.8 The saphenous opening lies below and lateral to the pubic tubercle.

5.9 Skin of the sole of the foot is innervated by branches of the common peroneal nerve.

5.10 The lateral condyle of the femur is more prominent than the medial condyle.

5.11 The talus articulates with the navicular at a synovial joint.

5.12 The tibial nerve innervates all the muscles of the calf.

5.13 Lymph from deep inguinal nodes drains to external iliac nodes.

5.14 Eversion of the foot is associated with rotation of talus.

5.15 Popliteus is supplied by the deep peroneal nerve.

5.16 Adductor magnus has no attachment to the tibia.

5.17 Dorsiflexion at the ankle joint is associated with the spinal nerve S2.

5.18 Damage to the femoral nerve in the distal half of the thigh impairs active extension of the knee joint.

5.19 Soleus and gastrocnemius are supplied by branches of the common peroneal nerve.

5.20 Most venous return in the lower limbs occurs through deep veins.

5.21 The tibial nerve supplies the muscles of the extensor compartment of the leg.

5.22 Flexor hallucis longus is a powerful flexor of the ankle joint.

5.23 The femoral vein traverses the femoral canal.

5.24 When the calf muscle pump is active, valves in the perforating veins direct blood into the superficial veins.

5.25 The medial collateral ligament of the knee joint attaches to the capsule of the knee joint.

Each of the incomplete statements below is followed by four or five suggested answers or completions. Select the one that is most appropriate.

5.26 The muscles of the anterior compartment of the leg:
   a) include no evertors of the foot.
   b) receive some innervation from the femoral nerve.
   c) are not active during standing.
   d) include flexors of the ankle joint.
   e) are innervated by branches of the common peroneal nerve.

5.27 The following do not cross the ankle joint:
   a) tendo calcaneus.
   b) the short saphenous vein.
   c) extensor digitorum brevis.
   d) the sural nerve.

5.28 The following are true of tendo calcaneus, *except*:
    a) it receives attachment of soleus.
    b) it has a profuse blood supply.
    c) it is separated from skin by a bursa.
    d) it receives the medial head of gastrocnemius.

5.29 The iliotibial tract:
    a) has no attachment to the bony pelvis.
    b) receives the deep fibres of gluteus maximus.
    c) receives the attachment of tensor fasciae latae.
    d) enables gluteus medius to stabilise the knee joint.

5.30 The following are true of the hamstring muscles, *except*:
    a) they receive blood from the profunda femoris artery.
    b) they are supplied exclusively by the tibial part of the sciatic nerve.
    c) they often limit flexion of the hip joint when the knee joint is extended.
    d) they are rotators of the flexed knee.

5.31 The femoral nerve:
    a) lies medial to the tendon of psoas major.
    b) arises from anterior divisions of the lumbar plexus.
    c) lies medial to the femoral artery.
    d) contains fibres from the ventral ramus of L4.

5.32 The superficial peroneal nerve supplies:
    a) extensor digitorum longus.
    b) flexor digitorum longus.
    c) peroneus longus.
    d) peroneus tertius.

5.33 Lymph drains to the superficial inguinal lymph nodes from the following, *except*:
    a) the femur.
    b) the great toe.
    c) the anterior abdominal wall below the umbilicus.
    d) the lower part of the anal canal.

5.34 Loss of sensation in the skin overlying the posterior aspect of the thigh would implicate:
   a) fibres of spinal nerve S2.
   b) the lateral cutaneous nerve of the thigh.
   c) the saphenous nerve.
   d) the sciatic nerve.

5.35 Near the ankle joint:
   a) flexor hallucis longus grooves sustentaculum tali.
   b) tibialis posterior grooves the lateral malleolus.
   c) the saphenous nerve lies posterior to the medial malleolus.
   d) the deep peroneal nerve supplies skin.

5.36 All the following are true if the sciatic nerve is divided, *except*:
   a) there is paralysis of the dorsiflexors of the foot.
   b) there is complete anaesthesia below the level of the knee.
   c) there is paralysis of the plantar flexors of the ankle.
   d) there is weakness of extension at the hip joint.

5.37 Which of the following does not articulate with talus?
   a) fibula.
   b) cuboid.
   c) tibia.
   d) calcaneus.

5.38 The following part of the femur is palpable:
   a) the shaft.
   b) the neck.
   c) the greater trochanter.
   d) the intercondylar notch.

5.39 The head of the femur:
   a) has no direct ligamentous attachment.
   b) articulates with the whole of the acetabulum.
   c) receives blood from branches of gluteal arteries.
   d) is partly covered by synovial membrane.

5.40 The dermatome L5 is associated with:
a) the penis.
b) the dorsum of the foot.
c) the medial side of the leg.
d) the anterior aspect of the thigh.
e) the little toe.

5.41 The following is true of the obturator nerve:
a) the main stem lies posterior to obturator externus.
b) the anterior division lies anterior to adductor longus.
c) it supplies sartorius.
d) it traverses the femoral canal.
e) it supplies adductor magnus.

5.42 Only nerves derived from the sciatic nerve supply sensory branches to:
a) the anterior aspect of the thigh.
b) the medial side of the leg.
c) the lateral surface of the leg.
d) the foot.
e) the knee joint.

5.43 While standing in the erect position the vertical line through the centre of gravity lies:
a) behind the coronal axis of the hip joint.
b) behind the coronal axis of the knee joint.
c) behind the coronal axis of the ankle joint.
d) parallel to the long axis of the femur.

5.44 Which of the following is not an antigravity muscle?
a) gluteus maximus.
b) quadriceps femoris.
c) gastrocnemius.
d) tibialis anterior.

5.45 The anterior cruciate ligament of the knee joint:
a) limits extension.
b) is intrasynovial.
c) has a good arterial supply.
d) resists posterior displacement of the tibia on the femur.

5.46 The following communicate with the knee joint:
   a) the prepatellar bursa.
   b) the infrapatellar bursa.
   c) the suprapatellar bursa.
   d) the semimembranosus bursa.

5.47 The superficial peroneal nerve supplies:
   a) peroneus tertius.
   b) the knee joint.
   c) skin of the sole of the foot.
   d) peroneus brevis.

5.48 The following are true of the capsule of the knee joint, *except*:
   a) it is lined by synovial membrane.
   b) it is attached to the articular margins of the femoral condyles.
   c) it is perforated by the tendon of popliteus.
   d) it is reinforced by an expansion from the tendon of semimembranosus.

5.49 Which of the following is not an abductor of the hip joint?
   a) gluteus maximus.
   b) gluteus medius.
   c) gluteus minimus.
   d) tensor fasciae latae.

5.50 The lumbosacral plexus gives origin to:
   a) posterior rami to muscles of the erector spinae group.
   b) anterior divisions to the femoral nerve which include fibres from L2, L3 and L4.
   c) the lumbosacral trunk to the lower limb.
   d) fibres from L5 to the anterior compartment of the leg.
   e) fibres from S1–S3 to the obturator nerve.

Each of the incomplete statements below is followed by five suggested answers or completions. Decide which are true and which are false.

5.51 The lateral collateral ligament of the ankle joint:
   a) attaches to talus.
   b) resists posterior displacement of the foot.
   c) attaches to the lateral malleolus.
   d) is crossed by peroneus brevis.
   e) attaches to calcaneus.

5.52 The tibial nerve:
   a) crosses the popliteal fossa.
   b) gives branches which supply skin of the dorsum of the foot.
   c) enters the posterior compartment of the leg by penetrating the inter-osseous membrane.
   d) is derived from the sciatic nerve.
   e) accompanies the posterior tibial artery.

5.53 Eversion of the foot:
   a) is less restricted than inversion.
   b) occurs mainly at the ankle joint.
   c) is produced by peroneus longus.
   d) is limited by the deltoid ligament.
   e) is produced by tibialis posterior.

5.54 The following muscles flex the knee joint:
   a) gastrocnemius.
   b) semitendinosus.
   c) adductor magnus.
   d) soleus.
   e) popliteus.

5.55 The patella:
   a) is covered by fibrocartilage on its posterior surface.
   b) tends to be displaced laterally by vastus lateralis.
   c) is a sesamoid bone.
   d) possesses a larger medial than lateral articular surface.
   e) is displaced proximally in relation to the tibial tubercle by contraction of quadriceps.

5.56 The great (long) saphenous vein:
   a) passes behind the medial malleolus.
   b) accompanies the saphenous nerve.
   c) communicates with the short saphenous vein via perforating veins.
   d) is compressed by the calf muscle pump.
   e) terminates in the femoral vein.

5.57 Branches of the femoral nerve supply:
   a) vastus medialis.
   b) pectineus.
   c) iliacus.
   d) skin of the foot.
   e) the ankle joint.

5.58 Tibialis anterior:
   a) has a tendon which is palpable.
   b) acts only on the ankle joint.
   c) is innervated by the superficial peroneal nerve.
   d) attaches to talus.
   e) lies medial to the anterior tibial artery.

5.59 Branches of the following nerves supply skin of the dorsum of the foot:
   a) the superficial peroneal nerve.
   b) the deep peroneal nerve.
   c) the sural nerve.
   d) the saphenous nerve.
   e) the tibial nerve.

5.60 The talus:
   a) receives the insertion of tendo calcaneus.
   b) gives attachment to the capsule of the ankle joint.
   c) supports body weight during standing.
   d) articulates with sustentaculum tali.
   e) gives attachment to the spring ligament.

5.61 The obturator nerve:
    a) arises from posterior divisions of the lumbar plexus.
    b) supplies adductor magnus.
    c) includes fibres from spinal nerve L3.
    d) supplies the knee joint.
    e) has no cutaneous branches.

5.62 The popliteal artery:
    a) is closely related to the knee joint.
    b) lies anterior to popliteus.
    c) lies posterior to the popliteal vein.
    d) gives origin to the peroneal artery.
    e) arises from the femoral artery.

5.63 Biceps femoris:
    a) attaches to the fibula.
    b) is innervated solely by the tibial component of the sciatic nerve.
    c) extends the hip joint.
    d) causes lateral rotation of the leg at the knee joint during the first few
       degrees of flexion at the knee.
    e) has a tendon which is closely related to the common peroneal nerve.

5.64 Gluteus maximus:
    a) contracts during changing to sitting from a standing position.
    b) contracts during changing to standing from a sitting position.
    c) stabilises the knee joint.
    d) abducts the hip joint.
    e) produces lateral rotation at the weight-bearing hip joint.

5.65 Lymph drains to superficial inguinal lymph nodes from:
    a) popliteal lymph nodes.
    b) the extensor compartment of the thigh.
    c) skin of the buttock.
    d) the foot.
    e) the penis.

5.66 Branches of the spinal nerve L3 are distributed to:
a) skin of the sole of the foot.
b) skin of the lateral side of the leg.
c) plantar flexors of the ankle joint.
d) skin over the anterior of the thigh.
e) extensors of the hip joint.

5.67 Which of the following muscles are usually innervated by branches from two different main nerve trunks?
a) adductor magnus.
b) vastus medialis.
c) biceps femoris.
d) triceps surae.
e) pectineus.

5.68 The following muscles adduct the hip joint:
a) sartorius.
b) gracilis.
c) semitendinosus.
d) quadratus femoris.
e) adductor magnus.

5.69 Quadriceps femoris:
a) controls flexion of the knee joint during movement from standing to squatting.
b) is a prime mover in extension of the knee joint.
c) assists flexion of the hip joint.
d) suffers complete paralysis if the femoral nerve is divided.
e) is supplied by fibres from spinal nerve L4.

5.70 The following contribute to stability of the knee joint:
a) synovial membrane.
b) quadriceps femoris.
c) the posterior cruciate ligament.
d) the suprapatellar pouch.
e) the medial collateral ligament.

5.71 The lateral meniscus:
    a) is larger in area than the medial meniscus.
    b) has a good arterial blood supply.
    c) is less mobile than the medial meniscus.
    d) alters its shape during flexion of the knee joint.
    e) receives an attachment from popliteus.

5.72 The synovial membrane of the hip joint:
    a) covers the pad of fat in the acetabular fossa.
    b) covers less of the femoral neck anteriorly than posteriorly.
    c) invests the ligament of the head of the femur.
    d) lines the inner surface of the capsule.
    e) is continuous with the psoas bursa.

5.73 During walking the following occur in the lower limb which bears weight:
    a) flexion at the knee joint.
    b) relaxation of quadriceps.
    c) extension at the hip joint.
    d) contraction of gastrocnemius.
    e) lateral rotation at the hip joint.

5.74 Which of the following muscles attach to the femur?
    a) semitendinosus.
    b) biceps femoris.
    c) gracilis.
    d) sartorius.
    e) rectus femoris.

5.75 The tibial nerve supplies:
    a) flexors of the hip joint.
    b) extensors of the hip joint.
    c) flexors of the knee joint.
    d) extensors of the knee joint.
    e) flexors of the ankle joint.

5.76 Fracture of the femur between the trochanters results in ischaemia of the head of the femur
because
most of the arterial blood to the femoral head passes through the neck from the femoral shaft.

5.77 Accumulation of fluid deep to the lower part of quadriceps femoris occurs after injury to the knee joint
because
the suprapatellar bursa communicates with the cavity of the knee joint.

5.78 Occlusion of the internal iliac artery results in ischaemia of the gluteal muscles on that side
because
there are no anastomoses between the gluteal arteries and branches derived from the external iliac artery.

5.79 Tibialis anterior is an evertor of the foot
because
its line of action lies medial to the axis of movement between the tarsal bones.

5.80 The muscle fibres of adductor magnus which attach to the ischial tuberosity are supplied by the medial (tibial) part of the sciatic nerve
because
that part of the muscle is formed from fibres which were associated with the hamstring group of muscles.

5.81 Occlusion of the distal part of the femoral artery causes ischaemia distal to the level of the knee joint
because
there are no anastomoses between branches of the femoral and popliteal arteries.

5.82 Fracture of the neck of the fibula may be associated with weakness of dorsiflexion at the ankle joint
because
fibres of the deep peroneal nerve may be injured near the neck of the fibula.

5.83 The patella tends to dislocate medially
because
the distal fibres of vastus medialis contract powerfully as the knee joint approaches full extension.

5.84 Discomfort may be felt along the medial side of the thigh in ovarian disease
because
pain involving the obturator nerve may be referred to the cutaneous distribution of its branches.

5.85 The hip joint is stable but allows little movement
because
many of the muscles which act on the hip joint are multipennate.

5.86 Incompetent valves in the perforating veins of the leg are associated with varicosities of the superficial veins of the legs
because
contraction of soleus forces venous blood proximally into the veins of the thigh.

5.87 Triceps surae is active during standing
because
ligaments do not contribute significantly to the stability of the ankle joint in a coronal plane.

5.88 Biceps femoris does not act on the knee joint
because
biceps has no attachments to the tibia.

5.89 Paralysis of quadratus femoris produces a marked limp
because
the only muscles which stabilise the extended knee joint are muscles which are supplied by the femoral nerve.

5.90 Division of the sciatic nerve results in some loss of sensation in the skin of the foot
because
the plantar nerves are derived from branches of the sciatic nerve.

5.91 An injection into the upper lateral quadrant of the buttock may cause damage to the sciatic nerve
because
in the buttock the sciatic nerve lies deep to gluteus maximus.

5.92 Vastus lateralis is a suitable muscle for intramuscular injections
because
no major vessels or nerves are at risk within the muscle.

5.93 A patient with damage to the second sacral nerve has 'foot drop'
because
the sciatic nerve includes fibres from spinal cord segments L4, L5, S1, S2 and S3.

5.94 Paralysis of the hamstring muscles is not serious
because
gluteus maximus can extend the hip joint and gastrocnemius can flex the knee joint.

5.95 The ankle joint is more stable in dorsiflexion than in plantar flexion
because
tension in the muscles of the anterior and posterior compartments of the leg stabilise the ankle joint.

5.96 When the knee joint approaches full extension, medial rotation of the femur in relation to the tibia occurs
because
semimembranosus and semitendinosus muscles are medial rotators at the knee joint.

5.97 Popliteus is important in extension of the knee joint
because
during the final stages of extension popliteus assists rotation at the knee to produce a stable 'locked' joint.

5.98 The ankle is an unstable joint
because
the muscles of the calf are weak compared to the muscles of the anterior compartment of the leg.

5.99 Posterior dislocation of the hip joint often damages the femoral artery
because
branches derived from the femoral artery supply the head of the femur.

5.100 The tendon of extensor hallucis longus is a useful guide when trying to palpate the pulse of the dorsalis pedis artery
because
the dorsalis pedis artery lies between the tendons of tibialis anterior and extensor hallucis longus.

5.101 In operations on the proximal part of the saphenous vein, the femoral vein is at risk of injury
because
the femoral vein does not lie within the femoral sheath.

In the following text, some words or phrases have been replaced by letters in brackets. Select the most appropriate word or phrase for each letter. Where a letter appears more than once, it represents exactly the same word or phrase.

5.102 The long saphenous vein arises from the (a)_____ side of the (b)_____ on the (c)_____ aspect of the foot. The vein passes (d)_____ to the (e)_____ malleolus and accompanies the (f)_____ nerve in the (g)_____ fascia overlying the (h)_____ aspect of the leg. At the level of the (i)_____ joint the vein lies (j)_____ to the joint and then crosses obliquely (k)_____ to the fascia lata overlying the (l)_____ compartment of the thigh. The long saphenous vein penetrates the (m)_____ fascia and terminates in the (n)_____ vein at the (o)_____ junction. During its course the long saphenous vein receives many (p)_____ and has communications with (q)_____ veins within the muscle compartments via (r)_____ veins which pierce the (s)_____ layer of the deep fascia. Changes in pressure cause blood to flow in directions determined by (t)_____ which permit blood in the (r) veins to flow from the (u)_____ veins into the (v)_____ veins.

5.103 The tibial nerve which originates from segmental (a)_____ nerves (b)_____ to (c)_____ usually separates from the (d)_____ nerve at the proximal angle of the (e)_____ fossa, but occasionally arises more proximally, in the (f)_____. Behind the knee joint the nerve lies (g)_____ to the (h)_____ artery and (i)_____ vein, lying deep to the fascial roof of the fossa. The tibial nerve receives (j)_____ fibres from the (k)_____ nerve which innervates skin on the (l)_____ aspect of the leg and (m)_____ side of the foot. The tibial nerve leaves the fossa between the two heads of (n)_____ and enters the (o)_____ compartment of the leg (p)_____ to the soleus muscle. In the leg the nerve is accompanied by the (q)_____ artery with its several (r)_____. The nerve supplies (s)_____ fibres to the muscles of the compartment, including those which (t)_____ the ankle joint, such as (n) and soleus which together attach via (u)_____, and those which (v)_____ the toes. At the ankle joint, the nerve lies (w)_____ to the (x)_____ malleolus and terminates as (y)_____ nerves which lie (z)_____ to the flexor retinaculum.

5.104 Identify the parts labelled a–l.

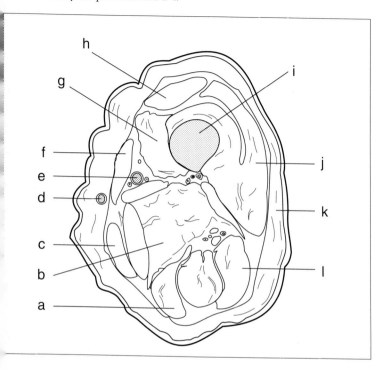

5.105 Pair each of the labelled parts a–l (see **Question 5.104**) with the most appropriate from the list of items A–L using each only once.

A stabilises the patella.
B flexes the hip joint.
C valves.
D 'tailor's muscle'.
E adipose tissue.
F nutrient artery.
G supplied by obturator and tibial nerves.
H hamstring muscle.
I in adductor canal.
J suitable for muscle biopsy.
K lateral rotator of flexed knee.
L most medial adductor.

5.106 Identify the parts labelled a–l.

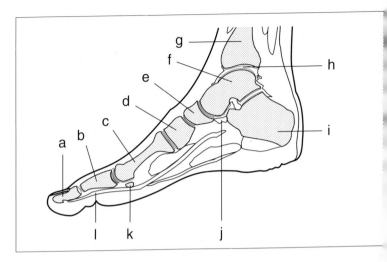

5.107 Pair each of the labelled parts a–l (see **Question 5.106**) with the most appropriate from the list of items A–L using each only once.

A attachment of Achilles tendon.

B muscle innervated by tibial nerve.

C subject to avascular necrosis if fractured.

D articulates with head of first metatarsal.

E attachment of flexor hallucis longus.

F attachment of peroneus longus tendon.

G articulates with distal phalanx.

H attachment of tibialis posterior tendon.

I 'march fracture'.

J supported by deltoid ligament.

K forms a fibrous joint with the fibula.

L forms a support for the head of talus.

6. HEAD AND NECK

Indicate whether the statement is true or false.

6.1 Geniohyoid is supplied from the ansa cervicalis.

6.2 Temporalis retracts the mandible.

6.3 The maxillary nerve supplies buccinator muscle.

6.4 In whispering the vocal folds are adducted but the arytenoids are separated.

6.5 The internal carotid artery traverses the foramen lacerum.

The incomplete statement below is followed by four suggested answers or completions. Select the one that is most appropriate.

6.6 The pharynx:
   a) is continuous with the oesophagus inferiorly.
   b) has a sensory supply from the hypoglossal nerve.
   c) receives motor fibres from the cranial root of the accessory nerve.
   d) has muscle fibres attached to the hyoid bone.

Each of the incomplete statements below is followed by five suggested answers or completions. Decide which are true and which are false.

6.7 The inferior alveolar nerve:
   a) is a branch of the maxillary division of the trigeminal nerve.
   b) conveys sensation from the skin of the lower lip.
   c) is closely related to the lingula of the mandible.
   d) innervates anterior belly of digastric muscle.
   e) terminates as the lingual nerve.

6.8 The left external carotid artery:
   a) has branches which convey blood to the eyeball.
   b) is related to the parotid gland.
   c) receives blood from the brachiocephalic artery.
   d) supplies the left lobe of the thyroid gland.
   e) is related to the hypoglossal nerve.

6.9 The internal carotid artery:
   a) arises from the brachiocephalic artery on the right.
   b) traverses the foramina transversaria of the cervical vertebrae 1–6.
   c) passes through the carotid canal.
   d) forms the basilar artery.
   e) gives rise to the anterior cerebral artery.

6.10 In the larynx:
   a) sensation from the vestibular fold travels in the superior laryngeal nerve.
   b) contraction of cricothyroid lowers pitch.
   c) the cricoid cartilage forms a complete ring.
   d) interarytenoid muscle adducts the vocal folds.
   e) the vocal fold is covered with respiratory epithelium.

6.11 The internal jugular vein:
   a) receives blood from the meninges.
   b) drains into the brachiocephalic vein.
   c) receives blood from the lingual vein.
   d) communicates with the retromandibular vein.
   e) has the ansa cervicalis on its anterolateral surface.

6.12 The head of the mandible:
   a) is moved medially by lateral and medial pterygoid acting together.
   b) is separated from the temporal bone by an intra-articular disc.
   c) is moved posteriorly by the lateral pterygoid muscle.
   d) is moved backwards by temporalis.
   e) most often dislocates backwards.

6.13 The intracranial dura mater:
a) is easily separated from the periosteum (endosteum).
b) forms the tentorium cerebelli.
c) contains the sigmoid sinus in the free edge of the falx cerebri.
d) is separated from the arachnoid mater by cerebrospinal fluid.
e) forms the diaphragma sellae.

6.14 The thyroid gland:
a) has an isthmus at the level of the second and third tracheal rings.
b) receives blood from a middle thyroid artery.
c) is enclosed in prevertebral fascia.
d) is closely related to both recurrent laryngeal nerves.
e) rises on swallowing.

6.15 The hypoglossal nerve:
a) leaves the skull through the jugular foramen.
b) conveys touch sensation from the anterior two-thirds of the tongue.
c) innervates the genioglossus muscle.
d) is closely related to the submandibular duct.
e) when damaged causes protrusion of the tongue to the damaged side.

6.16 The hypoglossal nerve:
a) supplies all the intrinsic muscles of the tongue.
b) leaves the skull through a canal (foramen) in the temporal bone.
c) lies between the internal carotid artery and internal jugular vein.
d) carries C1 fibres for the ansa cervicalis.
e) supplies mylohyoid.

6.17 Transection of the facial nerve in the internal acoustic meatus gives rise to:
a) paralysis of buccinator.
b) loss of taste in the anterior two-thirds of the tongue.
c) paralysis of masseter.
d) loss of secretion by the parotid gland.
e) paralysis of stapedius.

6.18 The maxillary air sinus:
- a) lies below the level of the hard palate.
- b) has a sensory supply from the ophthalmic division of the trigeminal nerve.
- c) is indented by the root of the canine tooth.
- d) is lined with respiratory epithelium.
- e) opens under the inferior concha.

6.19 Transection of the mandibular division of the trigeminal nerve in the foramen ovale gives rise to:
- a) paralysis of the buccinator muscle.
- b) loss of taste in the anterior two-thirds of the tongue.
- c) paralysis of the masseter muscle.
- d) loss of sensation from the inner surface of the cheek.
- e) paralysis of the mylohyoid muscle.

6.20 The maxillary division of the trigeminal nerve:
- a) is related to the cavernous venous sinus.
- b) conveys sensation from the lower eyelid.
- c) passes through the foramen rotundum.
- d) conveys sensation from part of the nasal septum.
- e) conveys sensation from the hard palate.

6.21 The tongue:
- a) has muscles supplied by the hypoglossal nerve.
- b) is attached to the mandible.
- c) is supplied with sensory fibres by the facial nerve.
- d) has two halves, separated by a midline septum.
- e) is comprised predominantly of smooth muscle.

6.22 The vagus nerve:
- a) passes through the jugular foramen.
- b) is contained in the carotid sheath.
- c) conveys touch sensation from the laryngopharynx.
- d) is motor to the stylopharyngeus muscle.
- e) carries fibres destined for the parotid gland.

6.23 In the neck:
   a) the phrenic nerve crosses anterior to the subclavian artery.
   b) the right vagus nerve gives rise to the recurrent laryngeal nerve.
   c) the scalenus anterior muscle passes posterior to the subclavian vein.
   d) the thoracic duct drains into the junction of the subclavian and internal jugular veins on the left.
   e) the sternohyoid muscle is supplied by the ansa cervicalis.

6.24 The glossopharyngeal nerve:
   a) passes through the jugular foramen.
   b) traverses the parotid gland.
   c) conveys touch sensation from the posterior third of the tongue.
   d) is motor to styloglossus.
   e) carries fibres destined for the submandibular gland.

6.25 In the nasal cavity and its walls:
   a) the nasolacrimal duct opens in the inferior meatus.
   b) the inferior concha is part of the ethmoid bone.
   c) the septum is supplied by the ophthalmic division of the trigeminal nerve.
   d) the vomer forms the posterior part of the septum.
   e) the sphenoidal air sinus opens into the middle meatus.

6.26 In the cranial cavity:
   a) the superior sagittal sinus drains into the transverse sinus.
   b) bleeding from the middle meningeal artery gives rise to extradural haematoma.
   c) the trigeminal ganglion is closely related to the apex of the petrous temporal bone.
   d) the internal carotid artery gives rise to the posterior cerebral artery.
   e) the temporal lobe lies in the anterior cranial fossa.

6.27 The mandible:
   a) is attached to the styloid process by a ligament.
   b) carries motor fibres to mentalis in the mandibular canal.
   c) is depressed by masseter.
   d) is elevated by lateral pterygoid.
   e) develops by ossification in membrane.

6.28 In the orbit:
   a) lateral rectus is supplied by the fourth cranial nerve.
   b) superior oblique produces adduction of the eye.
   c) the oculomotor nerve carries parasympathetic fibres to sphincter pupillae.
   d) secretomotor fibres for the lacrimal gland travel in the nasociliary nerve.
   e) the ophthalmic artery enters through the superior orbital fissure.

6.29 Sternocleidomastoid:
   a) attaches to the medial part of the clavicle.
   b) when contracted turns the face to the same side.
   c) is supplied by the cranial root of the accessory nerve.
   d) attaches to the mastoid process.
   e) is enclosed in the investing fascia of the neck.

6.30 The ophthalmic division of the trigeminal nerve:
   a) is closely related to the cavernous venous sinus.
   b) conveys sensation from the lower eyelid.
   c) passes through the foramen rotundum.
   d) conveys sensation from part of the nasal septum.
   e) conveys the sense of touch from the cornea.

6.31 The accessory nerve:
   a) passes through the jugular foramen.
   b) crosses the posterior triangle.
   c) conveys motor fibres to muscles of the soft palate.
   d) is motor to sternomastoid.
   e) carries fibres destined for the parotid gland.

6.32 Buccinator:
   a) is involved in drinking through a straw.
   b) attaches to the maxilla.
   c) is pierced by the submandibular duct.
   d) is active when pouting or pursing the lips.
   e) is supplied by the facial nerve.

6.33 The following muscles are attached to their paired bones:
a) masseter – zygomatic arch.
b) lateral pterygoid – condyloid process.
c) temporalis – temporal fossa.
d) medial pterygoid – lateral pterygoid plate.
e) levator veli palatini – scaphoid fossa of sphenoid.

6.34 Transection of the oculomotor nerve in the superior orbital fissure gives rise to:
a) paralysis of medial rectus.
b) loss of accommodation.
c) paralysis of superior oblique.
d) loss of secretion from the lacrimal gland.
e) paralysis of inferior oblique.

6.35 The vertebral artery:
a) arises from the brachiocephalic artery on the right.
b) traverses the foramina transversaria of the cervical vertebrae 1–6.
c) passes through the condylar foramen.
d) forms the basilar artery .
e) gives rise to the posterior inferior cerebellar artery.

6.36 The following muscles receive motor innervation from the paired nerve:
a) masseter – mandibular division of trigeminal nerve.
b) lateral rectus – oculomotor nerve.
c) stylopharyngeus – glossopharyngeal nerve.
d) cricothyroid – recurrent laryngeal nerve.
e) tensor veli palatini – mandibular division of trigeminal nerve.

6.37 The maxilla:
a) has the lateral pterygoid muscle attached to its tubercle.
b) contains the lesser palatine foramen.
c) forms the anterior bony boundary of the infratemporal fossa.
d) contributes to the hard palate.
e) forms part of the floor of the orbit.

6.38 In the mouth:
   a) the parotid duct opens opposite the first lower molar tooth.
   b) the palatoglossal ridge lies anterior to the palatine tonsil.
   c) the facial nerve conveys sensation from the tongue.
   d) the foramen caecum represents the origin of the pituitary gland.
   e) the submandibular duct crosses superior to the lingual nerve.

6.39 The recurrent laryngeal nerve:
   a) is a branch of the vagus nerve.
   b) conveys sensation from the laryngeal inlet.
   c) is closely related to the parathyroid glands.
   d) innervates cricothyroid muscle.
   e) supplies the inferior constrictor muscle.

6.40 The parotid gland:
   a) lies anterior to the ramus of the mandible.
   b) is traversed by the glossopharyngeal nerve.
   c) drains its secretions into the mouth alongside the frenulum of the tongue.
   d) is supplied by parasympathetic fibres from the facial nerve.
   e) contains lymph nodes.

6.41 Concerning dural venous sinuses:
   a) they lie between dura mater and arachnoid mater.
   b) the cavernous sinus surrounds part of the internal carotid artery.
   c) the sigmoid sinus drains through the foramen magnum.
   d) the superior sagittal sinus receives cerebrospinal fluid.
   e) the inferior sagittal sinus drains into the straight sinus.

6.42 Concerning the temporomandibular joint:
   a) the mandibular condyle is spherical in shape.
   b) the temporomandibular joint is divided into two by a disc composed of hyaline cartilage.
   c) depression of the mandible is caused by the action of the lateral pterygoid muscles.
   d) the muscles acting at the temporomandibular joint contain both fast and slow myosins.
   e) the temporomandibular joint most often dislocates backwards.

6.43 The maxillary artery:
   a) arises in the parotid gland.
   b) supplies blood to the lower teeth.
   c) supplies blood to the meninges of the middle cranial fossa.
   d) ends in the infratemporal fossa.
   e) supplies blood to the nasal cavity.

6.44 The vertebral artery:
   a) arises from the subclavian artery.
   b) traverses the foramina transversaria of the cervical vertebrae 1–6.
   c) passes through the condylar foramen.
   d) forms the circle of Willis with the internal carotid artery.
   e) gives rise to the posterior inferior cerebellar artery.

6.45 The soft palate:
   a) has sensation conveyed in the lesser palatine nerves.
   b) is attached anteriorly to the posterior edge of the hard palate.
   c) is elevated by palatopharyngeus.
   d) has respiratory epithelium on its inferior surface.
   e) is tensed by tensor veli palatini.

6.46 The facial artery:
   a) grooves the submandibular.
   b) passes deep to mylohyoid.
   c) is a direct branch of the external carotid artery.
   d) supplies the tongue.
   e) courses diagonally across the face to the inner angle of the eye.

Each question consists of two statements. Choose:
   a) if both statements are true and they are causally related.
   b) if both statements are true but they are not causally related.
   c) if the first statement is true and the second is false.
   d) if the first statement is false and the second is true.
   e) if both statements are false.

6.47 The thyroid gland rises during swallowing
   because
   the thyroglossal tract attaches to the tongue.

6.48 The tongue deviates to the uninjured side when protruded after hypoglossal nerve injury
because
the vertical intrinsic and genioglossal muscles are still able to contract on the uninjured side.

6.49 The retromandibular vein is particularly liable to damage in superficial parotidectomy
because
the facial nerve is deep in the parotid gland.

6.50 Infection may be carried from the superficial part of the face into the cranial cavity
because
facial veins communicate with the cavernous sinus via ophthalmic veins.

6.51 Bilateral superior laryngeal nerve palsy results in difficulty in breathing
because
cricothyroid muscle is supplied by the superior laryngeal nerve.

In the following text, some words or phrases have been replaced by letters in brackets. Select the most appropriate word or phrase for each letter. Where a letter appears more than once, it represents exactly the same word or phrase.

6.52 In the infratemporal fossa the lateral pterygoid muscle arises by two heads, the upper from (a)_____ and the lower from (b)_____. Its fibres pass backward to the (c)_____ and the (d)_____. Contraction of the muscle moves the head of the mandible (e)_____ and (f)_____ the mouth. Emerging from between the heads of the muscle is a sensory nerve, the (g)_____ from the (h)_____ division of the (i)_____ division of the (j)_____ nerve. Emerging from the lower border of the muscle and directed anteromedially is the (k)_____, joined by the (l)_____ posteriorly. Close by and directed anterolaterally is the (m)_____ which gives off the (n)_____ before entering the mandible adjacent to the (o)_____. Crossing lateral to all these structures is the (p)_____.

6.53 The pharynx possesses a lumen divided into (a)_____ superiorly, (b)_____ inferiorly and (c)_____ between them. The anterior boundary of (a) is formed by the posterior border of (d)_____ whilst inferiorly it is limited by the (e)_____. In the lateral wall of (a) is the opening of the (f)_____, whose posterior margin bears the (g)_____, from which extends (h)_____ muscle. Inferior to (f) is a bulge containing the (i)_____ muscle. Closely associated with (f) is a collection of lymphatic tissue, the (j)_____. In the roof of (a) is another lymphoid collection, the (k)_____ and in the (c) is a third lymphoid collection, the (l)_____. Lateral to (l) is the (m)_____ muscle and the (n)_____ artery.

6.54 Identify the indicated structures in this coronal section of the head.

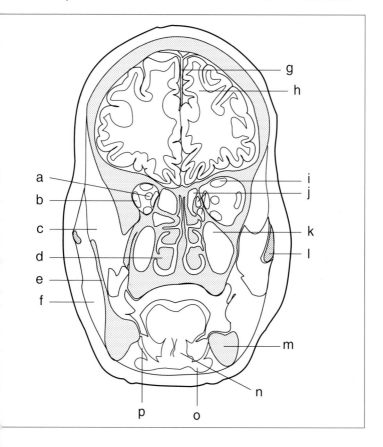

7. BACK

7.1 A sample of CSF may be obtained by inserting a needle into the subarachnoid space.

7.2 Paraplegia due to ischaemic damage to the spinal cord is a complication of operations on the abdominal aorta in which lumbar arteries are sacrificed.

7.3 Spinal dura mater extends inferiorly to vertebral level S2.

7.4 The spinous process of 'vertebra prominens' is easily palpable because during development of the secondary cervical curvature there is some posterior displacement of the vertebral body and hence its arch and processes.

7.5 The inferior articular facets of vertebra L5 face more anteriorly than those of the other lumbar vertebrae which face more laterally.

7.6 When bending forwards whilst standing, erector spinae controls flexion of the lumbar vertebral column.

7.7 Anterior displacement of vertebra L5 in relation to the sacrum causes compression of the sacral segments of the spinal cord.

7.8 Rotation occurs in both the cervical and thoracic regions of the vertebral column.

7.9 The sacrum is crossed by the obturator nerve.

7.10 Reduced volume of the intervertebral discs is a major cause of loss of height in elderly ladies.

7.11 The lumbosacral trunk is an anterior relation of the sacroiliac joint.

7.12 The vertebral canal contains the dorsal root ganglia.

7.13 The spinal cord terminates at the level of the intervertebral disc between vertebrae L1 and L2.

Each of the incomplete statements below is followed by five suggested answers or completions. Select the one that is most appropriate.

7.14 Movement of the bony pelvis during walking and/or running normally involves all of the following, *except*:
a) lateral flexion at the lumbar intervertebral joints.
b) movement at the hip joints.
c) contraction of the gluteal muscles.
d) movement at the sacroiliac joints.
e) rotation of the thoracic vertebral column.

7.15 The vertebral canal:
a) is of uniform calibre throughout its length.
b) contains the anterior longitudinal ligament.
c) terminates at the vertebral level L1 or L2.
d) is bordered partly by ligamenta flava.
e) contains a plexus of valved veins.

7.16 The spinal cord:
a) gives segmental branches from the filum terminale.
b) gives direct attachment to the spinal nerves.
c) would not be damaged by lumbar puncture between vertebrae L1 and L2.
d) gives rise to three pairs of sacral spinal nerves.
e) contains a space filled with CSF.

7.17 The following are true of the joints between the atlas and the axis, *except*:
a) they are all synovial joints.
b) they allow rotation of the head.
c) they are supported by the alar ligaments.
d) dislocation is usually fatal.
e) they are closely related to the first cervical nerves.

7.18 Which of the following is not true of the components of the erector spinae group of muscles?
a) deep muscles are generally longer than more superficial muscles.
b) they attach to the sacrum.
c) they are supplied segmentally.
d) they are enclosed by thoraco-lumbar fascia.
e) they are important postural muscles.

7.19 A typical thoracic vertebra:
a) has a body which contains inactive (fatty) bone marrow.
b) has synovial joints with adjacent vertebrae.
c) possesses costovertebral joints which are easily dislocated.
d) has joints which prevent rotation.
e) has no obvious basivertebral foramen.

7.20 The following are true of the sacrum, *except*:
a) forms part of the bony pelvis.
b) has a sacral canal which contains cauda equina.
c) articulates with the ilium at a cartilaginous joint.
d) is broader anteriorly than posteriorly.
e) provides attachment for the anterior longitudinal ligament.

7.21 The erector spinae group of muscles includes:
a) transversospinalis.
b) multifidus.
c) longissimus.
d) semispinalis.
e) splenius.

7.22 The axis vertebra:
a) has direct attachment to the skull.
b) lacks foramina transversaria.
c) is devoid of muscle attachment.
d) articulates with adjacent vertebrae only by synovial joints.
e) possesses no spinous process.

7.23 Which of the following is not true of the epidural space?
a) it contains fat.
b) it separates the periostium of the vertebral canal from the dura.
c) it is a site chosen for instillation of anaesthetics.
d) it extends into the sacral canal.
e) it extends into the cranial cavity.

7.24 The following are true of intervertebral discs, *except*:
a) they form special cartilaginous joints.
b) each is closely related to two spinal nerves.
c) each has a peripheral anulus fibrosus consisting of laminae of elastic fibres.
d) disc prolapse involves herniation of the nucleus pulposus.
e) when disc prolapse occurs the spinal nerve is usually compressed in the intervertebral foramen.

7.25 The vertebral column:
a) has eight cervical vertebrae.
b) is laterally flexed by one erector spinae muscle acting unilaterally.
c) allows minimal rotation in the cervical region.
d) has movements restricted by muscles rather than ligaments.
e) is rotated to the left by the left external abdominal oblique muscle.

Each of the incomplete statements below is followed by four or five suggested answers or completions. Decide which are true and which are false.

7.26 The sacroiliac joint:
a) is a fibrous joint.
b) allows rotation in which the lower part of the sacrum moves anteriorly.
c) is stabilised by the sacrotuberous ligament.
d) is a posterior relation of the common iliac artery.
e) is stabilised by the iliolumbar ligament.

7.27 The lumbar region of the vertebral column:
a) is the least mobile region of the spinal column.
b) has vertebrae which possess mamillary processes.
c) provides attachment for psoas major muscle.
d) has a secondary curvature.
e) has spinous processes which overlap the body of the adjacent vertebra below.

7.28 The spinal cord:
a) gives rise to eight pairs of cervical spinal nerves.
b) is continuous above with the medulla oblongata.
c) is intimately related to the pia mater.
d) extends throughout the length of the vertebral canal.
e) is separated from CSF by the arachnoid mater.

7.29 The intervertebral disc between vertebrae L3 and L4:
a) is the smallest intervertebral disc.
b) is related anteriorly to the abdominal aorta.
c) provides attachment for psoas major.
d) is related posteriorly to the cauda equina.
e) is closely related to the second lumbar nerves.

7.30 The atlas vertebra:
a) has no spinous process.
b) possesses no foramina transversaria.
c) has a synovial joint between its posterior arch and the odontoid process.
d) permits nodding movements at its joints with the cranium.
e) is closely related to the vertebral arteries which cross anterior to the atlanto-axial joints.

7.31 Intervertebral discs:
a) occur between the bodies of adjacent vertebrae.
b) are intimately related to the anterior and posterior longitudinal ligaments.
c) are secondary cartilaginous joints.
d) form part of the borders of the intervertebral foramina.
e) contribute to the curvatures of the vertebral column.

7.32 The sacrum:
a) usually consists of fused components of five vertebrae.
b) is concave anteriorly.
c) is attached by ligaments to the ilium and ischium.
d) forms synovial joints with the articular processes of the fifth lumbar vertebra.
e) possesses on each side an ala which is crossed by the anterior ramus of the fifth lumbar nerve.

7.33 The thoracic vertebral column:
a) has a primary curvature which is concave anteriorly.
b) includes 12 vertebrae.
c) has articular processes which lie on a near-sagittal plane.
d) throughout its length is related anteriorly to the oesophagus.
e) permits lateral flexion which is restricted by the ribs.

7.34 The thoracolumbar fascia:
 a) provides attachment for transversus abdominis muscle.
 b) encloses psoas major muscle.
 c) has no attachment to bone.
 d) provides attachment for external abdominal oblique muscle.

7.35 The spinal dura mater:
 a) is a thin tenuous layer.
 b) lies deep to the spinal arteries.
 c) lies superficial to the subdural space which contains the vertebral plexus of veins.
 d) terminates at the level of vertebra L2.
 e) attaches to the periostium of the foramen magnum.

7.36 A typical cervical vertebra possesses:
 a) a pair of synovial facet joints.
 b) a bifid spinous process.
 c) a relatively small vertebral foramen in relation to the spinal cord.
 d) a relatively small body.
 e) foramina transversaria.

7.37 The vertebral canal:
 a) transmits the vertebral artery.
 b) is fully developed at birth.
 c) contains a venous plexus.
 d) is narrowed by rotation of the head.
 e) lies anterolateral to the cervical vertebral discs.

7.38 The following relate to lumbar puncture at vertebral level L4/L5:
 a) the spinal cord terminates at a higher vertebral level.
 b) at this level the ligaments are thinner and easier to penetrate.
 c) there is no overlap of the spinous processes.
 d) the spinal nerves which exit from the vertebral column below this level are of little importance.
 e) the subarachnoid space extends inferiorly to below vertebral level L5.

7.39 The joints between vertebrae C4 and C5 allow:
 a) rotation.
 b) flexion with rotation.
 c) pure extension.
 d) lateral flexion.
 e) lateral flexion with rotation.

7.40 Tumours may spread to the bodies of the vertebrae via the vertebral plexus of veins and its communications with pelvic plexuses of veins
because
these plexuses do not contain valves and blood flow is determined by pressure changes in the veins.

7.41 Following a lumbar puncture, headache due to leakage of CSF into the epidural space occurs
because
the removal of CSF lowers the pressure in the subarachnoid space.

7.42 Stability at the individual joints between adjacent cervical vertebrae is not important
because
in the cervical region the vertebral canal is relatively large in relation to the spinal cord.

7.43 Lateral flexion of the cervical vertebral column produces simultaneous rotation between the vertebrae
because
the sternomastoid and trapezius muscles which produce lateral flexion are also rotators of the neck.

7.44 Knowledge of the segmental innervation of muscles and skin in the limbs is not helpful in the diagnosis of intervertebral disc disease
because
most muscles and areas of skin are innervated by neurons from more than one segment of the spinal cord.

7.45 Herniation of the nucleus pulposus of the intervertebral disc between lumbar vertebrae 3 and 4 commonly involves the fourth lumbar nerve
because
the fourth lumbar nerve passes between vertebrae L3 and L4.

7.46 Paralysis of both upper limbs follows fracture dislocation of the spine at the midthoracic level
because
in the thoracic region the arterial supply to the spinal cord is reinforced by branches from the intercostal arteries.

7.47 Instability between vertebrae L5 and S1 is likely to be associated with anterior displacement of vertebra L5
because
the intervertebral disc between L5 and S1 is wedge shaped.

7.48 Rectus abdominis muscle produces flexion of the lumbar vertebral column
because
the muscle is innervated by branches of ventral rami of spinal nerves.

7.49 An archeologist was right to say about an isolated human vertebra with a short spinous process that it must be lumbar'
because
all the cervical vertebrae possess bifid spinous processes.

7.50 During lumbar puncture, palpation of the iliac crests is not helpful
because
the vertebral level of the iliac crests varies considerably between individuals.

7.51 Little rotation occurs in the lumbar region of the vertebral column
because
the main muscles which produce rotation of the trunk do not act on the lumbar region of the vertebral column.

In the following text, some words or phrases have been replaced by letters in brackets. Select the most appropriate word or phrase for each letter. Where a letter appears more than once, it represents exactly the same word or phrase.

7.52 External to the spinal dura mater but within the (a)_____ canal lies the (b)_____ space. The space contains (c)_____ tissue and many veins which form the (d)_____ plexus. Superiorly the veins communicate with intracranial (e)_____ via the foramen (f)_____. The plexus receives (g)_____ from the spinal cord and from vertebral bodies via the (h)_____ veins. The plexus communicates with veins in front of the vertebral bodies which form the (i)_____ plexus and with the (j)_____ plexus formed by veins within the muscles posterior to the vertebral (k)_____. There are (l)_____ linking the three plexuses and the segmental veins of the thorax, abdomen and pelvis, namely the (m)_____, (n)_____ and (o)_____ veins respectively. Neither these veins nor the plexuses possess (p)_____ so that the (q)_____ of flow of blood depends on pressure. During coughing or straining when pressure in the (r)_____ and (s)_____ cavities is (t)_____, venous blood is squeezed (u)_____ the vertebral system of veins so that disease from pelvic organs, for example tumour of the (v)_____ in men, may spread readily to the (w)_____.

7.53 Identify the parts labelled a–g.

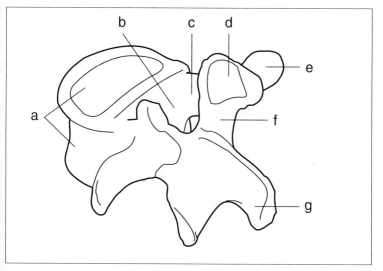

7.54 Pair each of the labelled parts a–g (see **Question 7.53**) with the most appropriate from the list of items A–G using each only once.

A synovial joint.
B intervertebral notch.
C palpable.
D ligamentum flavum.
E cauda equina.
F red bone marrow.
G rotatores.

7.55 Identify the parts labelled a–l.

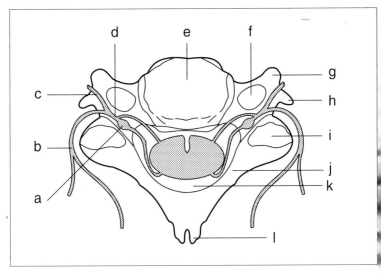

7.56 Pair each of the labelled parts a–l (see **Question 7.55**) with the most appropriate from the list of items A–L using each only once.

A vertebral artery.

B anterior longitudinal ligament.

C scalenus anterior.

D nerve cell bodies.

E cervical plexus.

F ligamentum nuchae.

G mixed nerve.

H supplies extensor muscles of neck.

I spinal cord.

J scalenus medius.

K synovial joint.

L laminectomy.

1. THORAX

T = true; F = false

1.1	T
1.2	F – the left pulmonary artery is connected to aortic arch via ligamentum arteriosum
1.3	T
1.4	T
1.5	F – chordae tendineae are attached to atrioventricular valves
1.6	d – crista terminalis is on the anterior wall of right atrium
1.7	a – the vein passes in front of the artery
1.8	c – the left 2nd and 3rd passes through left superior intercostal to left brachiocephalic vein; the right 2nd and 3rd pass through right superior intercostal vein to azygos vein
1.9	b – the right atrium
1.10	a T; b T; c T; d T; e F – parietal pleura is in contact with deep surface of rib
1.11	a F – two lobes; b T; c T; d F – lung is in contact with pericardium; e T
1.12	a T; b F – innervated by vagus nerve; c T; d F – related to left recurrent laryngeal nerve; e F – in contact with posterior surface of trachea
1.13	a T; b F – medial third of clavicle; c F – xiphisternum; d T; e F – 5th left intercostal space midclavicular line
1.14	a F – transverse process, tubercle of rib; b F – cartilaginous joint; c F – below; d F – cervical vertebrae have foramina transversaria; e F – obliquely posteriorly and downwards
1.15	a T; b F – right ventricle is part of inferior border; c T; d T; e T
1.16	a F – left brachiocephalic vein; b T; c T; d F – sensory supply from recurrent laryngeal nerves; e T
1.17	a
1.18	b – chordae tendineae and papillary muscles prevent eversion of valve cusps during systole, valves open passively during diastole

1.19   e – particles are more likely to enter the lower lobe of the right lung because the right main bronchus is more vertical

1.20   b – carcinoma spreads from one breast to the other via superficial lymphatics

1.21   d – thoracic duct drains into junction of left internal jugular and subclavian veins

1.22   a

1.23   a – biopsy is performed at full expiration as recess has greatest extent

1.24   e – only the left lung will collapse because there is no communication between the two spaces

1.25 a) ascending aorta/root of aorta
     b) aortic
     c) anterior interventricular artery
     d) anterior interventricular
     e) circumflex artery
     f) atrioventricular
     g) coronary sinus
     h) great cardiac vein
     j) posterior vein of left ventricle
     k) right atrium
     l) right
     m) right
     n) right

1.26 a) posterior or dorsal root
     b) posterior or dorsal root ganglion
     c) anterior or ventral root
     d) posterior ramus
     e) anterior ramus
     f) white ramus communicans
     g) unmyelinated
     h) grey ramus communicans
     j) dermatome
     k) myotome

1.27 a) cervical spinal nerves 3, 4 and 5
  b) inlet
  c) vein
  d) mediastinal pleura
  e) right brachiocephalic vein
  f) superior vena cava
  g) right atrium
  h) left common carotid artery
  j) aortic arch
  k) pulmonary trunk
  l) left atrial appendage
  m) left ventricle
  n) mediastinal and diaphragmatic
  o) pericardium and diaphragmatic peritoneum

1.28 a) sinuatrial node
  b) anterior
  c) right
  d) superior vena cava
  e) crista terminalis
  f) atrioventricular node
  g) interatrial
  h) coronary sinus
  j) atrioventricular bundle of His
  k) bundle branch
  l) moderator band

1.29 a) right atrium
  b) right ventricle
  c) left atrium
  d) left ventricle
  e) right lung
  f) oesophagus
  g) vertebral canal containing spinal cord
  h) vertebral body
  i) descending thoracic aorta

2. UPPER LIMB

2.1   F – extensor pollicis brevis attaches to the proximal phalanx of the thumb

2.2   T

2.3   T

2.4   F – the axillary nerve is accompanied by the posterior circumflex humeral artery, the profunda brachii artery by the radial nerve

2.5   T – scapular movement, which depends on the clavicular joints, is essential for full mobility of the upper limb

2.6   F – the axillary nerve passes posteriorly above and the radial nerve below teres major

2.7   F – the muscle is supplied from the brachial plexus via the thoracodorsal nerve

2.8   T – when the scapula is fixed contraction of pectoralis minor assists inspiration

2.9   F – intercostal nerves and posterior rami of thoracic nerves supply skin overlying latissimus dorsi

2.10   F – the clavicle is a bone of the upper limb, not part of the axial skeleton

2.11   T

2.12   T – the clavicular head of pectoralis major is visible and palpable when flexing the shoulder joint

2.13   F – the flexor retinaculum limits the carpal tunnel anteriorly

2.14   T – palmaris longus, acting through the palmar aponeurosis, may be regarded as a vestigial flexor muscle of the metacarpo-phalangeal joints

2.15   T – although the relationships between artery and nerve vary at different levels in the limb

2.16   T – the tendon within the joint is covered by synovial membrane which forms a bursa deep to the transverse humeral ligament

2.17   T – the distal fibres attached to the capsule used to be called 'articularis cubiti'

2.18	T – although brachioradialis is a flexor of the elbow joint it is supplied by the nerve of the extensor compartments
2.19	F – the median nerve provides the main supply to the thumb
2.20	T – the muscle is supplied by branches of the median (via anterior interosseous) and ulnar nerves
2.21	F – the long thoracic nerve to serratus anterior arises from roots C5, C6 and C7 of the brachial plexus
2.22	F – the ulnar nerve is the continuation of the medial cord which arises from the lower trunk
2.23	T
2.24	F – trapezius does not receive its motor supply via the brachial plexus but by the spinal accessory nerve
2.25	T
2.26	b – latissimus dorsi, teres major and pectoralis major are medial rotators whereas supraspinatus has no rotatory action
2.27	b – the mnemonic PAD may help you to remember that the Palmar interossei ADduct
2.28	c – abductor pollicis brevis attaches to the proximal phalanx of the thumb
2.29	a – scaphoid is also palpable anteriorly distal to the radius
2.30	c – extensor pollicis longus attaches to the ulna and interosseous membrane
2.31	e – the clavicle is commonly fractured by falls
2.32	e
2.33	e – extensor digitorum (and extensor indicis) are innervated by branches of the radial nerve
2.34	b – biceps spans the humerus and acts on the shoulder, elbow and radioulnar joints
2.35	c
2.36	d
2.37	c – the median nerve is subject to compression within the carpal tunnel in the 'carpal tunnel syndrome'

2.38   a – dermatome T1 lies on medial side of arm and forearm

2.39   c – the lateral cutaneous nerve of the forearm arises from the musculocutaneous nerve, a branch of the lateral cord

2.40   d – the anterior interosseous nerve does not supply deep extensors, arises from the median nerve, and contains sensory (proprioceptive and pain) fibres

2.41   b – spinal nerves C8 and T1 supply the intrinsic muscles of the hand

2.42   c – extensor carpi radialis longus is innervated by a branch which arises from the radial nerve proximal to the elbow joint

2.43   a – although c, d and e contribute to stability the rotator cuff is of primary importance

2.44   c – synovial sheaths for flexor tendons of the thumb and little finger pass between the digit and the forearm

2.45   c – the nerve to flexor carpi ulnaris arises in the forearm: the median nerve in addition to the ulnar supplies lumbrical and thenar muscles

2.46   a – the median nerve lies deep to the palmar aponeurosis. The palmar cutaneous branch (of the median nerve) which supplies skin of the palm lies superficial to the aponeurosis

2.47   d – adductor pollicis is usually supplied by the ulnar nerve alone

2.48   b

2.49   c – the brachial plexus includes fibres which traverse the anterior rami of spinal nerves (and also postganglionic sympathetic fibres)

2.50   a F; b T; c T; d T; e F

2.51   a F; b T; c T; d F; e T. Opponens pollicis and, paradoxically, the first dorsal interosseous act on the carpometacarpal joint of the thumb

2.52   a F; b T; c F; d T; e F

2.53   a F; b T; c T; d T; e T. Deltoid is active during abduction and also adduction when its fibres progressively relax to control adduction produced by gravity

2.54   a T; b F; c F; d T; e T. a, d and e pass between the clavicle and pectoralis minor

2.55   a T; b T; c T; d T; e F. Trapezius rotates and stabilises the scapula

2.56   a T; b T; c F; d F; e F. Neither sternomastoid nor pectoralis major elevate the scapula, the main component of 'shrugging the shoulders'

2.57   a F; b F; c F; d F. The movements could be produced by other muscles not supplied by the musculocutaneous nerve

2.58   a T; b T; c T; d T; e F. Rhomboid major assists rotation of the scapula during adduction

2.59   a T; b T; c T; d F; e F

2.60   a F; b F; c T; d T; e T. The anterior fibres of deltoid produce medial rotation, the posterior fibres lateral rotation

2.61   a T; b T; c F; d F; e T. The cephalic vein lies superficial to the roof of the fossa

2.62   a T; b T; c T; d T; e T. Anastomoses between collaterals help maintain perfusion, particularly if there is progressive occlusion of an artery

2.63   a F; b T; c F; d T; e F

2.64   a T; b F; c T; d T; e F. The axillary nerve is derived from the posterior cord, the ulnar nerve from the medial cord

2.65   a T; b T; c T; d F; e T. Brachioradialis flexes the elbow joint

2.66   a F; b F; c F; d F; e F. The long thoracic nerve is a branch from the roots: all the other nerves arise from the posterior cord

2.67   a T; b T; c F; d F; e F. The radial collateral ligament attaches to the anular ligament but does not stabilise the radioulnar joint

2.68   a T; b F; c T; d F; e T. Superficial muscles attach to the deep surface of the investing deep fascia

2.69   a F; b T; c T; d F; e F. The long thoracic nerve (of Bell) arises from roots C5, C6 and C7 of the brachial plexus

2.70   a F; b F; c F; d F; e F. All except brachialis attach to two of the three bones

2.71   a T; b T; c T; d T; e T. The skin is supplied by nerves whose territories overlap

2.72   a T; b F; c T; d F; e F. The thumb gains its mobility including opposition through the first carpometacarpal joint

2.73   a T; b T; c T; d T; e T. Most lymph traversing the axilla drains via the apical nodes into the subclavian lymph trunk

2.74   a T; b T; c T; d F; e F. Assymetry would result from weakness or wasting of trapezius (due to damage to a), serratus anterior (b), deltoid (c)

2.75   b – skin at the insertion of deltoid is supplied by the upper lateral cutaneous nerve, a branch of the axillary nerve

2.76   a

2.77   d – the main extensors of the shoulder joint are latissimus dorsi and deltoid

2.78   c – distally biceps attaches to the tuberosity of the radius and the subcutaneous border of the ulna

2.79   b – deltoid is an abductor because its line of action passes superior to the axis of movement

2.80   d – grip is impaired by radial nerve injury because the carpal extensors are weakened

2.81   d – everyday activities are minimally affected by absence of pectoralis major

2.82   e – the proximal part is abducted by supraspinatus

2.83   e – skin creases tend to lie parallel to the axis of movement

2.84   c – skin of the palm is innervated by palmar cutaneous branches (of the median and ulnar nerves) which cross anterior to the flexor retinaculum

2.85   a

2.86   d – abductor pollicis is the muscle to examine in 'carpal tunnel syndrome'

2.87   d – the main stem of the median nerve supplies flexor digitorum superficialis. Forearm flexor muscles are supplied by branches of both the median and ulnar nerves

2.88   a – the triangular cartilage separates the distal radioulnar joint from the radiocarpal (wrist) joint

2.89   e – flexor digitorum profundus attaches to the distal phalanges

2.90   a

2.91   b – pronator teres is not a powerful flexor because its line of action lies close to the axis of movement

2.92   e – the tendon of abductor pollicis longus can be palpated as the muscle belly contracts during extension of the thumb

2.93   c – the lateral cord contains fibres from C5 and C6, the medial cord C8 and T1

2.94   d

2.95   b – the musculocutaneous nerve supplies biceps which is a powerful supinator

2.96   e – the fibres in the membrane pass inferiorly and medially

2.97   c – brachioradialis assists rotation of the forearm to a position midway between the extremes of supination and pronation

2.98   b

2.99   c – trapezium and trapezoid are carpal bones on the lateral aspect of the distal row which articulate with metacarpal bones

2.100 a) anterior
     b) arm
     c) two heads
     d) short head
     e) long head
     f) coracobrachialis
     g) supraglenoid
     h) deep
     i) transverse humeral
     j) synovial membrane
     k) glenohumeral
     l) subcutaneous or medial
     m) aponeurosis
     n) cubital
     o) tuberosity of the radius
     p) musculocutaneous
     q) elbow
     r) shoulder
     s) radioulnar
     t) supinator
     u) flexion

2.101 a) lateral
b) venous arch
c) dorsal
d) 'snuff box'
e) superficial radial
f) elbow
g) anterolateral
h) basilic
i) superficial
j) biceps brachii
k) deltoid
l) pectoralis major
m) inferior
n) axillary
o) lymphatics
p) lateral
q) infraclavicular
r) axilla

2.102 a) abductor pollicis longus
b) extensor indicis
c) extensor pollicis brevis
d) dorsal tubercle of radius (Lister)
e) tendon of extensor carpi ulnaris
f) tendon of extensor carpi radialis brevis
g) tendon of extensor carpi radialis longus
h) tendon of extensor pollicis longus
i) tendon of extensor digitorum

2.103 a) E
      b) I
      c) B
      d) D
      e) H
      f) C
      g) F
      h) A
      i) G

2.104 a) blade of scapula
      b) subscapularis
      c) serratus anterior
      d) axillary fat
      e) pectoralis major
      f) head of humerus
      g) glenoid fossa
      h) deltoid
      i) infraspinatus

2.105 a) C
      b) E
      c) D
      d) B
      e) F
      f) I
      g) A
      h) G
      i) H

3. ABDOMEN

3.1   T

3.2   F – enters the second part

3.3   F – lies posterior to left kidney

3.4   T

3.5   T

3.6   F – cystic duct joins common hepatic duct to form the bile duct

3.7   F – taeniae coli meet at base of appendix

3.8   d – in contact with right dome at bare area

3.9   c – psoas major is medial

3.10   a T; b F – ninth costal cartilage; c F – anterior surface of costal margin; d T; e T

3.11   a F – ninth costal cartilage; b T; c T; d T; e T

3.12   a F – superior mesenteric; b F – intraperitoneal; c T; d T; e F – medial wall

3.13   a T; b F – coeliac artery; c F – most of duodenum is retroperitoneal; d T; e T

3.14   a F; b T; c T; d F; e T

3.15   a F – lies to the left of the midline; b T; c F – passes into pelvis to supply rectum; d F – middle colic is a branch of superior mesenteric; e T

3.16   a

3.17   d – renal veins drain into systemic venous system ie inferior vena cava

3.18   a

3.19 a) coeliac axis or artery
    b) coeliac plexus
    c) thoracic splanchnic nerves
    d) sympathetic trunks
    e) preganglionic
    f) left gastric artery
    g) oesophageal
    h) splenic artery
    j) short gastric
    k) gastroepiploic artery
    l) spleen

3.20 a) fundus
    b) gas or air
    c) cardia or cardiac orifice
    d) lesser curvature
    e) left gastric
    f) hepatic portal vein
    g) oesophageal
    h) porta-caval anastomosis
    j) oesophageal varices
    k) anterior vagal trunk
    l) left

3.21 a) costal margin
    b) splenic flexure of colon
    c) fourth part of duodenum
    d) superior mesenteric vessels
    e) head of pancreas
    f) second part of duodenum
    g) liver
    h) spleen
    i) left kidney
    j) psoas major
    k) aorta
    l) right renal vein joining IVC

4. PELVIS AND PERINEUM

4.1   F – the external meatus is narrower

4.2   F – to left renal vein on left and inferior vena cava on right

4.3   T

4.4   T

4.5   T

4.6   F – posterior

4.7   d – sympathetic for ejaculation

4.8   b – inferior surface

4.9   a – in superficial pouch

4.10   c – parasympathetic from S2, 3 and 4

4.11   c – the superficial fascia of the thigh is bound to the deep fascia just distal to the inguinal ligament

4.12   a T; b T; c F – most inferior between uterus and rectum; d T; e T

4.13   a T; b F – uterine artery from internal iliac; c F – rectum; d T; e F – to internal iliac and aortic nodes

4.14   a F – direct to para-aortic nodes; b F – testicular artery direct from aorta; c T; d F – dartos muscle in scrotal wall controls distance of testis from body and thus temperature; e T

4.15   a T; b F – fundus is more mobile; c T; d T; e F – has a supravaginal portion

4.16   a T; b T; c F – does not leave the pelvic cavity; d T; e T

4.17   a T; b T; c F – there is no peritoneum in contact with the prostate; d F – posterior; e T

4.18   a T; b F – posterior; c F; d T; e T

4.19   a

4.20   c – levator ani contracts during coughing and sneezing

4.21   d – cervix uteri can be palpated per rectum

4.22   a

4.23   e – the perineal branch is given off after the ischial spine

4.24   a

4.25   e – the bladder enlarges superiorly because the trigone is posteroinferior and fixed.

4.26 a) pubic/pubis
   b) tendinous arch
   c) obturator/pelvic
   d) obturator internus
   e) ischial spine
   f) downwards, medially and posteriorly
   g) perineal branch of S4
   h) pudendal nerve
   i) elevation
   j) increased angulation
   k) ischiorectal or ischioanal fossa

4.27 a) external iliac artery
   b) internal iliac artery
   c) ureter
   d) ovary
   e) umbilical/superior vesical
   f) obturator artery
   g) internal pudendal artery
   h) middle rectal artery
   i) uterine
   j) vaginal
   k) superior gluteal artery
   l) greater sciatic foramen
   m) inferior gluteal artery

4.28 a) common iliac artery
   b) ovarian vessels
   c) ureter
   d) internal iliac artery
   e) uterine tube
   f) superior vesical artery
   g) obturator artery
   h) ovary
   i) uterine artery
   j) fundus of uterus
   k) rectouterine pouch
   l) lumen of bladder
   m) vagina

5. LOWER LIMB

5.1   F – with the exception of vastus medialis, contraction of the parts of quadriceps tends to displace the patella laterally

5.2   F – the knee joint is stable in full extension without contraction of quadriceps which is relaxed

5.3   F – the short saphenous vein usually perforates the popliteal fascia to terminate in the popliteal vein

5.4   T – the ischial fibres of adductor magnus lie behind the axis of movement at the hip joint

5.5   T

5.6   F – fibres from the tibial nerve contribute to the innervation of adductor magnus

5.7   T – in addition, L3 supplies skin over the medial side of the thigh

5.8   T – digital pressure here will occlude the saphenofemoral junction

5.9   F – the sole is innervated by branches of the tibial nerve

5.10   T

5.11   T – the talonavicular joint is a synovial ball and socket joint

5.12   T

5.13   T

5.14   F – the talus is anchored in the mortice of the ankle joint: the other tarsal bones rotate about talus

5.15   F – the tibial nerve supplies popliteus

5.16   T – the distal attachment is to the femur

5.17   F – the segmental nerves associated with dorsiflexion are L4 and L5

5.18   F – the femoral nerve divides into its terminal branches in the proximal part of the thigh

5.19   F – the tibial nerve supplies gastrocnemius and soleus

5.20   T

5.21   F – the common peroneal nerve supplies the anterior (extensor) compartment of the leg

5.22   F – flexor hallucis longus lies close to the axis of movement at the ankle joint so that it is a weak flexor

5.23   F – the femoral vein lies immediately lateral to the femoral canal within the femoral sheath

5.24   F – the valves direct blood from superficial into deep veins within the muscle compartments

5.25   T – the posterior aspect of the ligament blends with the capsule of the knee joint

5.26   e

5.27   c – extensor digitorum brevis is confined to the dorsum of the foot

5.28   b

5.29   c – the superficial part of gluteus maximus stabilises the knee joint via the iliotibial tract

5.30   b – the common peroneal nerve supplies the short head of biceps femoris

5.31   d – the femoral nerve arises from posterior divisions of the lumbar plexus, itself formed from ventral rami of spinal nerves

5.32   c – peroneus tertius and extensor digitorum longus are innervated by the deep peroneal nerve

5.33   a – lymph arising deep to the investing fascia does not drain to superficially located lymph nodes

5.34   a – the posterior cutaneous nerve of thigh distributes fibres from S2 to skin over the back of the thigh

5.35   a

5.36   b – the saphenous nerve is a branch of the femoral nerve

5.37   b – cuboid articulates with the navicular (fibrous joint) and calcaneus (synovial joint) but not with talus

5.38   c

5.39   c – the gluteal arteries contribute to the cruciate and trochanteric anastomoses from which blood enters retinacular vessels which supply the femoral head

5.40   b

5.41   e – adductor magnus is innervated by the obturator nerve and the tibial part of the sciatic nerve

5.42   c – branches of the femoral, obturator and sciatic nerves supply the lower limb

5.43   a – the hip joint is stabilised in extension by the iliofemoral ligament which reduces the muscular activity required while standing

5.44   d

5.45   a – the cruciate ligaments are intracapsular but extrasynovial and have a limited arterial supply

5.46   c – the suprapatella bursa may be regarded as part of the joint: the semimembranosus bursa sometimes communicates with the knee joint

5.47   d – peroneus tertius is supplied by the deep peroneal nerve

5.48   b – capsular attachments do not follow the articular margins anteriorly, laterally or posteriorly

5.49   a – although the upper part of gluteus maximus does abduct, the muscle is primarily an extensor of the hip joint

5.50   d – the main segments supplying the anterior compartment of the leg are L4 and L5

5.51   a T; b T; c T; d T; e T. The calcaneofibular ligament lies medial to peroneus brevis

5.52   a T; b F; c F; d T; e T

5.53   a F; b F; c T; d F; e F

5.54   a T; b T; c F; d F; e T. Gastrocnemius is a weak flexor: the main role of popliteus is to 'unlock' the knee joint at the start of flexion

5.55   a F; b T; c T; d F; e F

5.56   a F; b T; c F; d F; e T. Perforating veins penetrate investing fascia and link superficial with deep veins

5.57   a T; b T; c T; d T; e F. Pectineus may also receive innervation from the obturator nerve

5.58   a T; b F; c F; d F; e T. Talus is devoid of tendon and muscle attachments - one of the reasons it is vulnerable to avascular necrosis after fracture

5.59   a T; b T; c T; d T; e T. All except the saphenous nerve are derived from the sciatic nerve

5.60   a F; b T; c T; d T; e F. The 'spring ligament' is the plantar calcaneonavicular ligament

5.61   a F; b T; c T; d T; e F. The obturator nerve arises from anterior divisions of the lumbar plexus, the femoral nerve from posterior divisions.

5.62   a T; b F; c F; d F; e T. Relationships in the popliteal fossa may be confused because the fossa is usually examined from behind

5.63   a T; b F; c T; d F; e T. Biceps laterally rotates the knee flexed at about 90°: the tibial nerve innervates the long head, the common peroneal nerve the short head

5.64   a T; b T; c T; d T; e F. Gluteus maximus is an important antigravity muscle at the hip joint and stabilises the knee joint through the iliotibial tract

5.65   a F; b F; c T; d T; e T. Some but not all lymph from the foot and penis drains to these nodes

5.66   a F; b F; c F; d T; e F. Areas of skin innervated by segmental nerves are: sole, L5 and S1; lateral leg, L5; hip extensors, L5, S1

5.67   a T; b F; c T; d F; e T

5.68   a F; b T; c F; d T; e T. Sartorius is a weak abductor of the hip joint

5.69   a T; b T; c T; d T; e T. During squatting from a standing position quadriceps progressively relaxes to allow controlled flexion at the knee joint

5.70   a F; b T; c T; d F; e T. Quadriceps femoris is crucial to stability of the knee joint

5.71   a T; b F; c F; d T; e T. The menisci have a relatively meagre arterial blood supply

5.72   a T; b F; c T; d T; e T. Anteriorly the femoral neck is entirely intracapsular and covered by synovial membrane

5.73 a F; b F; c T; d T; e F. Medial rotation occurs at the hip joint of the weight-bearing limb and increases the length of the stride

5.74 a F; b T; c F; d F; e F. All these muscles act on both the hip and knee joints: the short head of biceps attaches to the femur

5.75 a F; b T; c T; d F; e T. Flexion at the hip joint and extension at the knee joint are controlled primarily by branches of the femoral nerve

5.76 e – intertrochanteric fractures do not usually result in ischaemia of the femoral head

5.77 a – the suprapatellar bursa may be emptied by compression to assist diagnosis of knee injuries

5.78 c – occlusion of the internal iliac arteries may also lead to failure of penile erection

5.79 d – the axis of movement in inversion passes through the head and neck of talus

5.80 a

5.81 c – anastomoses may sustain arterial blood flow which is adequate at rest but inadequate during exercise which leads to intermittent claudication

5.82 a

5.83 d

5.84 a – referred pain may be due to irritation of pelvic peritoneum or of the nerve itself

5.85 e

5.86 b – valves in the veins of the leg direct blood flow from superficial to deep and from distal to proximal

5.87 c – stability in a sagittal plane during standing requires activity in the leg muscles

5.88 d

5.89 e – quadriceps femoris, not quadratus femoris: the gluteal nerves supply muscles which stabilise the knee via the iliotibial tract

5.90 a – the saphenous nerve (from the femoral nerve) supplies skin on the medial aspect of the foot and would be intact in sciatic nerve injury

5.91 d – the upper lateral quadrant is the 'safe area' away from sciatic nerve: injections here are deposited in gluteus medius and minimus

5.92 a

5.93 d – the main spinal nerves associated with dorsiflexion are L4 and L5

5.94 d – the hamstring muscles are essential in walking, running and climbing

5.95 b

5.96 b – the rotation as the joint approaches full extension is secondary to ligament attachments and the shape of the articular surfaces

5.97 e

5.98 e – the calf muscles are large and powerful

5.99 d – posterior dislocation of the hip joint may damage the sciatic nerve

5.100 c – the dorsalis pedis artery lies lateral to the tendon of extensor hallucis longus

5.101 c – the femoral vein is at risk during dissection of tributaries near the saphenofemoral junction

5.102 a) medial
b) venous arch
c) dorsal
d) anterior
e) medial
f) saphenous
g) superficial
h) anteromedial
i) knee
j) posteromedial
k) superficial
l) anterior
m) cribriform
n) femoral
o) saphenofemoral
p) tributaries
q) deep
r) perforating
s) investing

t) valves
u) superficial
v) deep

5.103 a) spinal
b) L4
c) S3
d) sciatic/common perineal
e) popliteal
f) buttock
g) superficial
h) popliteal
i) poplitea
j) sensory
k) sural
l) posterolateral
m) lateral
n) gastrocnemius
o) posterior/flexor
p) anterior/deep
q) posterior tibial
r) venae comitantes
s) motor/prorioceptive
t) plantar flex/flex
u) tendo calcaneus
v) flex
w) posterior
x) medial
y) medial and lateral plantar
z) deep

5.104 a) semimembranosus
b) adductor magnus
c) gracilis
d) great saphenous vein
e) femoral artery
f) sartorius
g) vastus medialis
h) rectus femoris
i) shaft of femur
j) vastus lateralis
k) superficial fascia
l) biceps femoris

5.105 a) H
b) G
c) L
d) C
e) I
f) D
g) A
h) B
i) F
j) J
k) E
l) K

5.106 a) distal phalanx of hallux
b) proximal phalanx of great toe
c) first metatarsal
d) medial cuneiform
e) navicular
f) talus
g) tibia
h) ankle joint
i) calcaneum
j) calcaneonavicular ('spring') ligament
k) sesamoid bone
l) tendon of flexor hallucis longus

5.107 a) E
b) G
c) I
d) F
e) H
f) C
g) K
h) J
i) A
j) L
k) D
l) B

6. HEAD AND NECK

6.1   F – geniohyoid by C1 fibres in hypoglossal nerve

6.2   T

6.3   F – facial nerve supplies buccinator

6.4   T

6.5   F – internal carotid is associated with the superior part of the foramen lacerum

6.6   b – hypoglossal only motor; Vc, IX and X supply sensation to pharynx

6.7   a F – branch of mandibular division; b T; c T; d T; e F – terminates as mental nerve

6.8   a F – blood supply to eye from internal carotid; b T; c F – receives blood from common carotid and aortic arch; d T; e T

6.9   a F – branch of common carotid; b F – vertebral artery traverses vertebrae; c T; d F – vertebral artery forms basilar; e T

6.10   a T; b F – cricothyroid raises pitch; c T; d T; e F – vocal fold is stratified squamous epithelium

6.11   a T; b T; c T; d T; e T

6.12   a T; b T; c F – moved anteriorly by lateral pterygoid; d T; e F – usually dislocates anteriorly

6.13   a F – very tightly bound to endosteum; b T; c F – inferior sagittal sinus in free edge of falx cerebri; d F – CSF is between arachnoid and pia; e T

6.14   a T; b F – middle thyroid artery does not exist; c F – pretracheal fascia; d T; e T

6.15   a F – hypoglossal canal; b F – motor nerve only, lingual touch for anterior $^2/_3$; c T; d F – lingual nerve closely related to submandibular duct; e T

6.16   a T; b F – hypoglossal canal is occipital bone; c T; d T; e F – Vc supplies mylohyoid

6.17   a T; b T; c F – Vc supplies masseter; d F – IX supplies parasympathetic to parotid gland; e T

6.18   a T; b F – maxillary division only; c T; d T; e F – opens under middle concha

6.19   a F – buccinator supplied by VII; b F – taste from anterior ²/₃ - chorda tympani - VII; c T; d T; e T

6.20   a T; b T; c T; d T; e T

6.21   a T; b T; c T; d T; e F – striated muscle

6.22   a T; b T; c T; d F – stylopharyngeus IX; e F – parotid IX

6.23   a T; b T; c T; d T; e T

6.24   a T; b F – VII traverses parotid gland; c T; d F – XII motor to styloglossus; e F – IX fibres - parotid gland

6.25   a T; b F – inferior concha is separate bone; c T; d T; e F – opens to sphenoethmoidal recess

6.26   a T; b T; c T; d F – posterior cerebral from basilar; e F – middle cranial fossa

6.27   a T; b F – mentalis VII; c F – masseter elevates mandible; d F – lateral pterygoid depresses mandible; e T.

6.28   a F VI; b F – abduction and depression; c T; d F – zygomatic nerve - lacrimal; eF – optic foramen.

6.29   a T; b F – opposite side; cF – spinal root; d T; e T.

6.30   a T; b F – maxillary division; c F – superior orbital fissure; d T; e T

6.31   a T; b T; c T; d T; e F – IX carries fibres for parotid gland

6.32   a T; b T; c F – pierced by parotid duct; d F – orbicularis oris; e T

6.33   a T; b T; c T; d T; e F – scaphoid fossa - tensor veli palatini

6.34   a T; b T; c F – superior oblique - IV; d F – VII - Vb - zygomatic nerve - lacrimal nerve; e T

6.35   a F – subclavian artery; b T; c F – foramen magnum; d T; e T

6.36   a T; b F VI; c T; d F – external laryngeal from superior laryngeal; e T

6.37   a F – medial pterygoid; b F – greater palatine foramen; c T; d T; e T

6.38   a F – second upper molar; b T; c T; d F – thyroid gland; e T

6.39   a T; b F – superior laryngeal nerve; c T; d F – superior laryngeal nerve; e T

6.40   a F – posterior to ramus; b F – facial nerve VII; c F – vestibule opposite second upper molar tooth; d F – IX; e T

6.41   a F – between layers of dura or dura and endosteum; b T; c F – jugular foramen; d T; e T

6.42   a F – condyloid; b F – fibrocartilage; c T; d T; e F – most often dislocates forwards

6.43   a T; b T; c T; d F – pterygomaxillary fissure; e T

6.44   a T; b T; c F – foramen magnum; d F – basilar - posterior cerebral; e T

6.45   a T; b T; c F – depresses soft palate; d F – stratified squamous; e T

6.46   a T; b F – lingual passes deep; c T; d F – lingual supplies tongue; e T

6.47   b – thyroid gland rises because it is held to thyroid cartilage by pretracheal fascia

6.48   d – tongue deviates to injured side

6.49   e – facial nerve is superficial to retromandibular vein, and only that part of the gland superficial to facial nerve is removed in superficial parotidectomy

6.50   a

6.51   d – bilateral recurrent laryngeal nerve palsy gives rise to breathing difficulty

6.52 a)   infratemporal surface of the greater wing of the sphenoid bone
    b)   lateral surface of the lateral pterygoid plate
    c)   neck of the mandible
    d)   capsule of the temporomandibular joint
    e)   forwards
    f)   opens
    g)   buccal nerve
    h)   anterior
    i)   mandibular
    j)   trigeminal V
    k)   lingual nerve

    l)   chorda tympani
    m) inferior alveolar nerve
    n)   nerve to mylohyoid
    o)   lingula
    p)   maxillary artery

6.53 a) nasopharynx
     b) laryngopharynx or hypopharynx
     c) oropharynx
     d) nasal septum, vomer
     e) raised soft palate
     f) auditory tube
     g) tubal elevation
     h) salpingopharyngeus
     i) levator veli palatini
     j) tubal tonsil
     k) pharyngeal tonsil/adenoid
     l) palatine tonsil
     m) superior constrictor
     n) facial

6.54 a) optic nerve
     b) lateral rectus
     c) temporalis
     d) inferior concha
     e) ramus of mandible
     f) masseter
     g) falx cerebri
     h) frontal lobe
     i) superior rectus
     j) ethmoid air cells
     k) maxillary air sinus/antrum
     l) zygomatic arch
     m) body of mandible
     n) genioglossus
     o) anterior belly of digastric
     p) mylohyoid

7. BACK

7.1   T

7.2   T – a recognised complication of aortic surgery

7.3   T

7.4   F – the spinous process is longer than adjacent processes

7.5   T

7.6   T – erector spinae may act as an antigravity muscle.

7.7   F – substantial displacement may cause compression of the cauda equina; cord ends at L2 vertebral level

7.8   T

7.9   T

7.10   F – collapse of vertebral bodies causes shortening

7.11   T

7.12   F – dorsal root ganglia lie in the intervertebral foramina

7.13   T – the conus lies at the level of the first lumbar intervertebral disc

7.14   d – minimal movement occurs at the sacroiliac joint

7.15   d – the veins lack valves

7.16   e – the central canal is continuous with the ventricular system

7.17   e – fracture dislocation between vertebrae C1 and C2 is 'hangman's fracture'

7.18   a

7.19   b – facet joints are synovial

7.20   c – the sacroiliac joint is synovial

7.21   c

7.22   a – apical and alar ligaments attach the dens to the occipital bone

7.23   e – extradural bleeding, usually arterial, may strip the dura from the bone of the skull

7.24 c – the fibres of the anulus are mainly collagen

7.25 b – there are eight cervical nerves but only seven cervical vertebrae

7.26 a F; b F; c T; d T; e T. The sacroiliac joint is synovial and supported by strong interosseous ligaments

7.27 a F; b T; c T; d T; e F. The lumbar column permits free flexion and extension

7.28 a T; b T; c T; d F; e F. The cord descends to vertebral level L2

7.29 a F; b T; c T; d T; e F. The nerve L2 passes above the third lumbar vertebra

7.30 a T; b F; c F; d T; e F. The dens articulates with the anterior arch of the atlas

7.31 a T; b T; c T; d T; e T

7.32 a T; b T; c T; d T; e T

7.33 a T; b T; c F; d F; e T. The aorta lies anterior to the lower thoracic vertebrae

7.34 a T; b F; c F; d F. External oblique has a free posterior margin

7.35 a F; b F; c F; d F; e T

7.36 a F; b T; c T; d T; e T. Two upper and two lower facet joints

7.37 a F; b F; c T; d T; e F. The vertebral foramen/canal should not be confused with the foramen transversarium

7.38 a T; b F; c T; d F; e T

7.39 a T; b T; c T; d T; e T. Rotation and lateral flexion always occur together

7.40 a

7.41 b – leakage of CSF is due to meningeal damage not to the transient lowering of CSF pressure

7.42 e – the spinal cord is particularly vulnerable to injury in the cervical region

7.43 b – the facet joints are inclined obliquely

7.44 d – the damaged intervertebral disc(s) may be identified using knowledge of segmental innervation

7.45   c – disc protrusion commonly involves the segmental nerve named one inferior to the disc; the fourth lumbar nerve passes between vertebrae L4 and L5

7.46   d – the upper limbs are innervated by the brachial plexus from spinal cord segments C5 to T1

7.47   b – the displacement is due to the inclination of the upper surface of the sacrum

7.48   b – flexion is produced because rectus abdominis lies anterior to the vertebral column

7.49   e – cervical vertebrae have short spinous processes and the atlas possesses a posterior tubercle

7.50   e – the iliac crests provide a reliable guide to vertebral level L3/L4

7.51   c – the abdominal muscles act on the lumbar vertebral column

7.52 a) vertebral
   b) epidural, extradural
   c) fatty, areolar
   d) vertebral
   e) venous sinuses
   f) magnum
   g) blood
   h) basivertebral
   i) anterior vertebral
   j) posterior vertebral
   k) arches, column
   l) anastomoses
  m) intercostal
   n) lumbar
   o) sacral
   p) valves
   q) direction
   r) thoracic
   s) abdominopelvic
   t) increased
   u) into
   v) prostate
   w) vertebrae

7.53 a) body of vertebra
b) vertebral foramen
c) pedicle
d) superior articular facet
e) transverse process
f) lamina
g) spinous process

7.54 a) F
b) E
c) B
d) A
e) G
f) D
g) C

7.55 a) dorsal root ganglion
b) posterior ramus
c) anterior ramus
d) spinal nerve
e) body of vertebra
f) foramen transversarium
g) anterior tubercle
h) posterior tubercle
i) articular facet
j) lamina
k) vertebral foramen
l) spinous process

7.56 a) D
b) H
c) E
d) G
e) B
f) A
g) C
h) J
i) K
j) L
k) I
l) F